# Explore
# Fraser Island

David & Julie Hinchliffe

Published by Great Sandy Publications
Feedback and comments welcome
Web: www.greatsandy.com.au
Post: PO Box 94, Robe SA,
Australia 5276
First edition May 2006
ISBN 0-9758190-0-3

Cover, words and photographs by
David & Julie Hinchliffe
Base map by Cottrell Cameron & Steen Surveys Pty Ltd

Printed in Australia by Fergies

Copyright maps, text and photographs: © Great Sandy Publications 2006. No part of this guide may be reproduced, transmitted, stored or retrieved by any means without the publisher's prior permission.

Your suggestions and updates are welcome. Note that conditions on Fraser Island are constantly changing. Beach and track access and conditions can vary, as can various features, services and regulations. Readers should source up-to-date information prior to travel.

*Explore Fraser Island* is intended as a guide only. Whilst every effort has been made to ensure the accuracy of information published herein, the publisher disclaims all responsibility and all liability for any problems arising from any errors, omissions, interpretations or applications of this guide book.

Photos: front cover (main) - Sandy Cape, inside photo - goat's foot flower, back cover (main) - Lake McKenzie.

## About the authors

David and Julie first discovered a shared fascination for natural areas while surveying sand dunes in southern Australia. After university, like many other graduates, the pair took jobs wholly unrelated to their studies. This landed them in the middle of the desert answering phones and carrying bags. Inspired by rumours of a spectacular sand island off Australia's east coast, they decided to pack the troopie and try for jobs as resort eco guides on Fraser Island. Luckily, the plan worked, giving them a unique chance to explore this amazing part of the world.

Since then, both Julie and David have pursued rewarding careers in conservation and land management. Julie has an Honours degree in coastal botany and is in demand as one of Australia's leading freelance environment writers. David has Honours and Masters degrees in environmental science and works as an ecologist on flora and habitat. They have two dogs - one, of course, named 'Fraser'.

## Acknowledgements

The authors are grateful to the many people who provided information and encouragement throughout the lengthy production of this guide. Particular thanks goes to family and friends for their support, Yumi Kenri for photographic equipment, Emma Parminter for keeping an eye on things, Dennis McCart for design advice, Kelly Birch for website design, Gail Lane for volunteering her editorial skills and Rose Darroch for the wallaby and flying fox photos.

# CONTENTS

| | |
|---|---|
| HOW TO USE THIS BOOK | 5 |
| FRASER ISLAND MAP | 6 |
| BASIC INFORMATION | 8 |
| Camping | 8 |
| Accommodation | 11 |
| Fishing | 12 |
| Environmental care | 13 |
| Bushwalking | 14 |
| Permits | 17 |
| Ferries & barges | 17 |
| Fuel & supplies | 18 |
| Phones & internet | 19 |
| What to take | 19 |
| Safety | 20 |
| Ranger stations | 21 |
| DRIVING ON FRASER ISLAND | 22 |
| Inland tracks | 23 |
| Beach access points | 24 |
| Beach driving | 25 |
| Creek crossings | 26 |
| Rocky outcrops | 27 |
| Western beach | 28 |
| Tyre pressure | 28 |
| Driving technique | 30 |
| Recovery techniques | 30 |
| Environmental driving | 34 |
| Quick tips | 35 |
| HISTORY | 36 |
| Aboriginal history | 36 |
| European history | 38 |
| World Heritage listing | 39 |
| FLORA & FAUNA | 41 |
| Island of birds | 41 |
| Dingo's domain | 44 |
| Dingo safety | 45 |
| Life in the dunes | 47 |
| Life in the lakes & swamps | 53 |
| Sand flat & shallow dwellers | 60 |
| Creatures of the deep | 62 |
| Animals to be wary of... | 64 |
| NORTHERN FRASER ISLAND | 68 |
| Northern Fraser Island Map | 69 |
| Sandy Cape | 70 |
| Sandy Cape Lighthouse | 72 |
| Rooney Point | 74 |
| Ocean Lake | 76 |
| Orchid Beach | 78 |
| Waddy Point | 80 |
| Champagne Pools | 82 |
| Indian Head | 84 |
| Wathumba Creek | 86 |
| CENTRAL FRASER ISLAND | 88 |
| Central Fraser Island Map | 89 |
| Awinya Creek | 90 |
| Lake Gnarann | 92 |
| White Lake | 94 |
| Lake Bowarrady | 96 |
| Bowarrady Creek | 98 |
| Woralie Creek | 100 |
| Red Canyon | 102 |
| Wungul Sandblow | 104 |
| Dundubara | 106 |
| Cathedral Beach | 108 |
| Lake Allom | 110 |
| Lake Coomboo | 112 |
| Boomerang Lakes | 114 |
| Moon Point | 116 |
| Knifeblade Sandblow | 118 |
| The Pinnacles | 120 |
| Maheno Wreck | 122 |
| Eli Creek | 124 |
| Yidney Scrub | 126 |
| Lake Garawongera | 128 |
| Happy Valley | 130 |
| Rainbow Gorge | 132 |
| Valley of the Giants | 134 |
| Kingfisher Bay | 136 |
| SOUTHERN FRASER ISLAND | 138 |
| Southern Fraser Island Map | 139 |
| Stonetool Sandblow | 140 |
| Lake Wabby | 142 |
| Eurong | 144 |
| Lake McKenzie | 146 |
| Basin Lake | 148 |
| Central Station | 150 |
| Pile Valley | 152 |
| Lake Jennings | 154 |
| Lake Birrabeen | 156 |
| Lake Benaroon | 158 |
| Lake Boomanjin | 160 |
| Dilli Village | 162 |
| Ungowa | 164 |
| Garry's Anchorage | 166 |
| Snout Point | 168 |
| Southern Track | 170 |
| Hook Point | 172 |
| FURTHER READING | 175 |

# Introduction

*Banksia flower*

# How to use this book

This guide book answers many of the questions travellers to Fraser Island ask time and time again.

We've aimed to keep the information as concise and relevant as possible, saving our personal opinions for the "what we think" sections. This means you can get a realistic idea of what to expect. As for which places to visit and routes to take - that's all up to you.

The different locations featured in this book reflect the diversity of places to see, tracks to take and experiences to be had on Fraser Island, but are by no means exhaustive. The more you see of Fraser Island, the more you'll realise just how diverse this incredible wilderness area is.

We've arbitrarily divided the island into three sections - Northern, Central and Southern - each with a corresponding map. Within each section, the locations are presented in clusters which roughly reflect their proximity to one another (rather than alphabetical order). Generally they appear in order from north to south.

Important information pertaining to the sub-headings is given below, although most are self-explanatory.

**Map reference**: Provides grid references for maps produced by Hema and Sunmap. These maps can be purchased from shops on the island as well as many on the mainland. While we have provided basic maps in this book to assist your planning, a more detailed one is definitely good to have on hand.

**Location**: Gives distances in kilometres from some of the locations featured in this book that you are likely to be driving or walking to or from. Stated distances are approximate only and may vary due to wheel slip in soft sand and / or reduced tyre pressures.

**What's there?** Lists some of the major features and attractions or 'draw cards'.

**Nearest to:** Lists the three nearest locations also featured in this book accessible by vehicle and / or on foot.

**Why go?** Describes features or activities that commonly attract visitors to the location or are of possible interest. May include facts about natural features, historical significance or events, unique attributes and, in some cases, the location's context in terms of the wider environment or global significance.

**Getting there**: Outlines the major access routes and describes general track and traffic conditions (not exhaustively since there are many tracks on the island and conditions and accessibility can vary). In some cases, walking routes are mentioned.

**Facilities**: Lists major visitor facilities such as parking, toilets, payphones, shops, picnic tables, campsites and others. Note facilities may have changed or improved since the time of writing.

**What to look for**: Describes features of interest to give you an idea of what you might encounter or experience at the location or on the way there. Includes interesting facts about natural or unusual formations, native wildlife and plants and different sights and sounds. Hopefully, this will inspire you and your companions to notice some of the amazing things around you.

**What we think**: Summarises our personal opinions about the location, including what we like or don't like about it and whether or not we think it's worth visiting. This may include tips on time frames, what to do or take with you and environmentally responsible behaviour.

Happy exploring...

# Introduction

# Fraser Island map

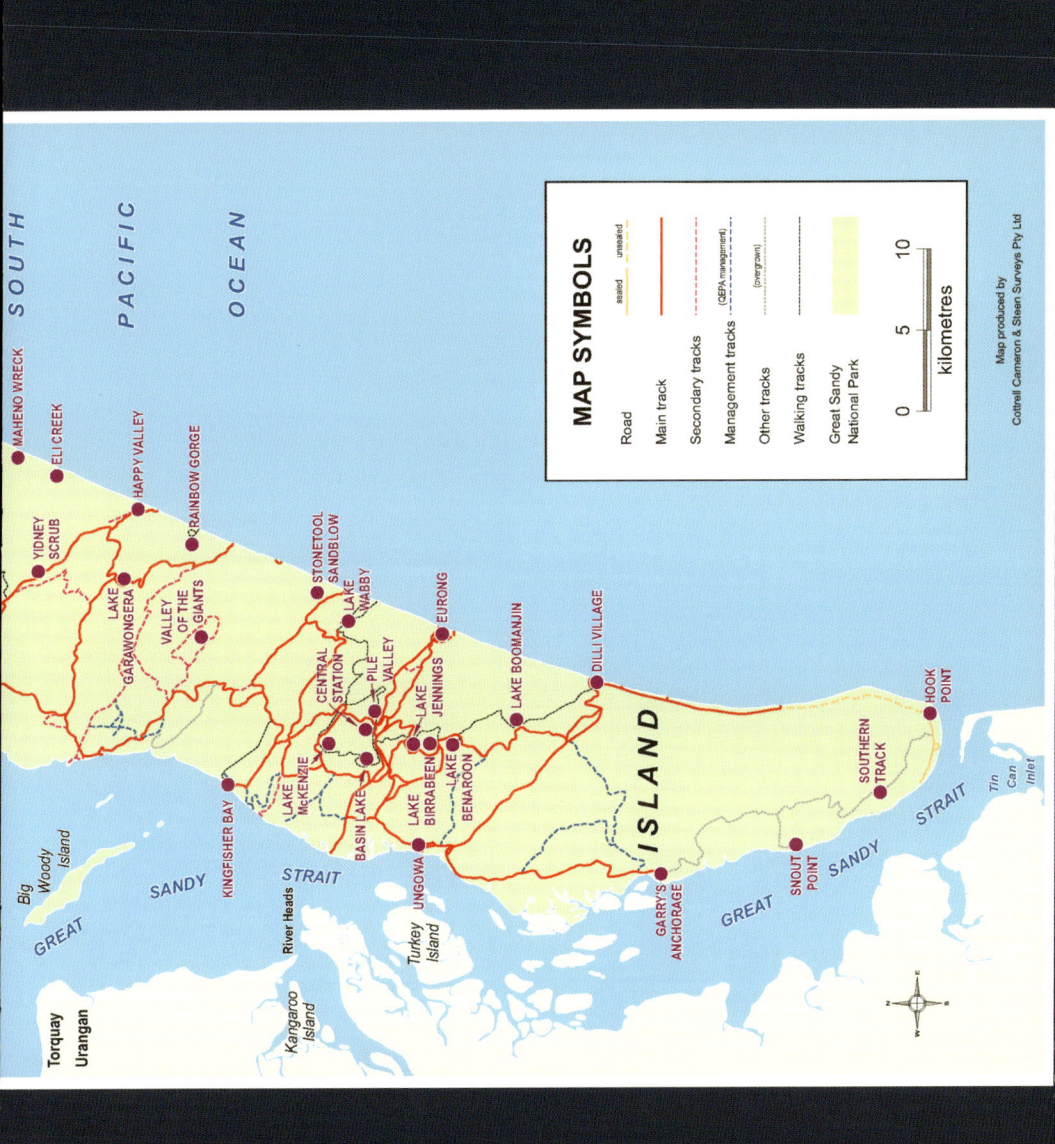

# Introduction

Fraser Island is well deserving of its World Heritage status. If you're into science - whether it's botany, geography or archaeology - you'll be amazed. The island overflows with natural, geological and cultural values. It is home to more than 350 species of birds, 900 species of plants and eight types of vegetation, ranging from pioneer dune communities to diverse wetlands, flowering heathlands and luxuriant rainforests - not bad for an island made entirely of sand.

Situated just off the coast of south east Queensland, Fraser Island extends for 123 kilometres, covering some 184,000 hectares. Together with the Cooloola sand mass to the south, the island forms part of the Great Sandy National Park.

Not only is Fraser Island the world's largest sand island, it supports some of the world's oldest dune systems. These are thought to be at least 750,000 years old. Dotted along its eastern shores are rainbow-coloured sand dunes, some carved into the shapes of cathedral-like spires. Further inland, the older, leached dunes and beaches appear pure white in stark contrast to the labyrinth of freshwater lakes that range from red to blue to green in colour.

K'gari, meaning "paradise", was the name given to the island by its traditional Aboriginal owners, the Butchulla. Their island home was rich in natural resources, including plenty of sources of fresh water and oceans stocked with an abundance of marine life. The Butchulla were a healthy, resourceful people, with an intimate knowledge of the island's natural cycles and special places. Over time, they had developed intricate methods for hunting and obtaining various plant and animal foods, fibres and medicines.

Today, Fraser Island has something for everyone whether its camping, fishing, swimming, hiking, whale watching, four-wheel driving or adventure seeking. It has to be one of the world's best places for getting up close and personal with nature.

But make sure you do some research before you go (this guide book is an excellent start). Despite its popularity, Fraser Island can be a relatively challenging and remote holiday destination. It's up to all who visit to enjoy the island in an informed, safe and environmentally-sensitive way, helping to protect its World Heritage values for generations to come.

## Camping

Camping is a great way to enjoy the island. Even during the busiest of times, it's always possible to find a relatively secluded campsite if you're prepared to bush camp. Those wanting some basic facilities are well catered for in the island's managed campgrounds. Whichever option you take, camping on the island of-

# Basic information

*Champagne Pools*

fers a chance to really get back to nature and enjoy a feeling of remoteness and escape. If you're quiet, you might spot some nocturnal wildlife and on a clear night, the stars can be spectacular.

A permit is required to camp anywhere within the national park, including Queensland Parks and Wildlife Service (QPWS)-managed campgrounds. About 98% of the island is national park. It is recommended you obtain your permit before leaving the mainland (see p17).

Those preferring to 'rough it' (free-range / bush camping) can set up camp along the eastern beaches just behind the small front dune (foredune). Obey all signs such as 'no camping' and 'closed for regeneration'. (Note Indian Head and surrounds are now permanently closed to camping.) You'll also need to camp at least 50 metres away from any freshwater creeks. Look for pre-existing clearings where others have camped before and only drive on pre-existing tracks, taking care not to damage the fragile dune vegetation.

A few beach camping options also exist on the island's west coast. Again, look for pre-existing sites, set up away from creeks and well clear of the high tide mark, and obey any signs. Biting insects can be particularly bothersome on the west coast; however many people are willing to put up with this in order to enjoy the wonderful scenery (Wathumba Creek being a good example).

Some of the west coast options accessible to drivers and / or hikers include Wathumba Creek, Garry's Anchorage, Awinya Creek, Bowarrady Creek, Woralie Creek, Moon Point, Urang Creek and Ungowa (this book includes detailed descriptions of all of these except Urang Creek). If you are unsure about the suitability of particular places for bush camping, check

9

# Introduction

*Campground amenities, Central Station*

with local rangers before setting up.

Bush campers should bury toilet waste to a depth of at least 50 cm, taking care to cover the pit before they leave. For privacy, many groups take a tent toilet with them, particularly during busy times. Don't attempt to bury rubbish - this should be taken with you and deposited in one of the bins provided elsewhere on the island. Remember to be courteous towards other campers. Most come to enjoy the peace and quiet.

Campfires have long been a favourite part of the camping experience, however open campfires are no longer permitted on Fraser Island. If you are keen to enjoy a campfire you will need to use one of the communal fire rings, so long as fire bans are not in place. QPWS provides these at Waddy Point beachfront campground, Waddy Point campground, and Dundubara campground. Since firewood is no longer provided by QPWS and campers are not permitted to collect their own wood from the forests (including fallen timber), you will need to bring your own clean supply of wood onto the island. This needs to be untreated, milled timber. Campfires should be extinguished using water before you leave the area and never left unattended. (For more information on the new campfire policy see the QPWS *Touring and camping guide for Fraser Island* which comes with your camping permit).

Although inland camping options are limited, a very extensive managed campground is provided at Central Station.

The island's main QPWS-managed campgrounds are located at:

- Central Station (fenced)
- Lake Boomanjin (fenced)
- Dundubara
- Waddy Point (upper area fenced)

Showers, picnic tables, barbecues, toilets and tap water are among the basic facilities provided. Groups or families concerned about dingoes often opt for the fenced campgrounds (see Dingo safety p 45). Central Station, Dundubara and Waddy Point have campground rangers.

Advanced bookings can only be made for Dundubara and Waddy Point (upper and lower campgrounds). During busy holiday periods - such as eastern state school holidays - you may need to book several months in advance. Phone 13 13 04 or visit www.epa.qld.gov.au and click on 'Camping bookings'. All other campsites located within the national park are available on a first come first served basis.

Other established QPWS campgrounds tend to receive less servicing and / or supervision. These include:

- Wathumba
- Lake McKenzie (hikers only)
- Ungowa (no showers)
- Garry's Anchorage (no showers)

# Basic information

Three privately-run campgrounds are available on the island. Camping at these locations is not covered by the QPWS camping fee. These are:

- Dilli Village (fenced). Now managed by The University of Sunshine Coast, sites can be booked by phoning (07) 4127 9130 (see p 162-3)
- Frasers at Cathedral Beach. Sites can be booked by phoning 1800 444 234 or 07 4127 9177 (see p 108-9)
- K'Gari Camping Area (see below)

The K'Gari Camping Area is located just inland, a few kilometres south of The Pinnacles (take the Woralie exit). This privately run alcohol-free campground, previously known as the Thoorgine Education and Cultural Centre, is now under new management. It provides campers with campsites, picnic tables, solar-powered hot showers, toilets and drinking water. There are also three large raised shelters, which provide cooking facilities (suitable for groups), as well as a permanent raised 'safari tent'. The campground is run by descendants of the island's Aboriginal people, the Butchulla.

For information on which facilities are available at the QPWS and other private campgrounds, refer to the location descriptions in this book.

## Accommodation

Visitors to Fraser Island can choose from a range of accommodation styles, varying from shacks and basic cabins through to luxurious villas, motel units and large privately-owned holiday houses.

For information on the many holiday house options on Fraser Island, talk to your travel agent or a local tourist information officer. Holiday houses are situated along the east coast at places like Eurong, Second Valley, Happy Valley, Indian Head and Orchid Beach.

Kingfisher Bay and Eurong support the largest resorts. Other major accommodation centres can be found at Happy Valley, Cathedral Beach and Dilli Village.

The main accommodation options are

Lake Boomanjin

# Introduction

described in some detail in the location descriptions throughout this book. Contact details in alphabetical order:

**Dilli Village**
Phone: (07) 4127 9130
Email: btaylor2@usc.edu.au
Web: www.usc.edu.au/Research/ResearchandInnovation/Facilities/FraserIsland/
For advance group bookings:
Phone: (07) 5430 2891
Email: sstephen@usc.edu.au

**Eurong Beach Resort**
Phone: (07) 4127 9122
Email: eurong@fraser-is.com
Web: www.fraser-is.com

**Frasers at Cathedral Beach**
Phone: 1800 063 933 or (07) 4125 3933
Email: tours@fraserislandco.com.au
Web: www.fraserislandco.com.au/frasers.html

**Fraser Island Wilderness Retreat (Happy Valley)**
Phone: 1800 063 933 or (07) 4125 3933
Email: tours@fraserislandco.com.au
Web: www.fraserislandco.com.au/retreat.html

**Kingfisher Bay Resort & Village**
Phone: 1800 072 555 or 07 4120 3333
Email: reservations@kingfisherbay.com
Web: www.kingfisherbay.com

**Orchid Beach Trading Post / Holiday Letting**
Phone: (07) 4127 9220
Web: www.orchidbeachtradingpost.net

## Fishing

Fraser Island has long been a drawcard for fishing enthusiasts, ranging from family groups who just want to throw in a few lines through to competitive sports anglers. As well as the opportunity for some good catches, part of the attraction is the unique natural beauty and sense of remoteness Fraser Island has to offer.

The variety of fish in the area reflects the island's location as well as the different offshore habitats it provides. Fraser Island falls within a region that is incredibly rich in fish species. The south east corner of Queensland is home to nearly one third of Australia's total number of marine and estuarine fish (well over 1000 species).

Just offshore, the variety of habitats is able to sustain a diverse mix of open water, rock and reef fish. Around the reef off Waddy Point, for instance, anglers can enjoy catches of red emperor, sweetlip, snapper, pearl perch, parrot fish and coral trout. Some people choose to fish off the rocks, while others launch small boats from the beach (although this is not recommended for the inexperienced).

Beach fishing is the most popular pursuit, with popular spots including Waddy Point, the Maheno, Moon Point, Orchid Beach, Sandy Cape and the calm side of Indian Head. The deep gutters along the eastern beach bring schools of fish close to shore to feed. By September, Fraser Island and Cooloola's surf beaches support annual spawning runs of tailor. This attractive silver-coloured fish, said to reach up to 14 kilograms in weight, was once a seasonally-significant food resource for the island's Butchulla people and visiting tribes. The tailor are known to congregate around the Maheno Wreck and Sandy Cape and when they're on the bite, will latch on to a simple silver lure or even a bare hook. Although not renowned as the best eating fish, they are exciting to catch

# Basic information

and best enjoyed when eaten fresh.

Other fish caught off the beaches include bream, tarwhine, whiting, tailor, dart (swallowtail), jewfish and flathead. Avoid fishing areas infested with fireweed (see p 65).

In good weather, private boats and fishing charters regularly visit the waters off Fraser. Those who come for sports fishing target species such as wahoo, jobfish and mackerel. In fact, off the northwest coast, mackerel and tuna can sometimes be seen schooling close to shore.

All up, thousands of anglers visit Fraser Island each year. Many come for the famous annual Toyota Fishing Classic held in the island's north-eastern corner (Waddy Point / Orchid Beach). This six day event takes place around the end of May to the start of June, attracting entrants interested in beach and reef fishing.

Fishing enthusiasts, like all other visitors to Fraser Island, need to be conscious of minimising their impact on the environment. Gone are the days when anglers would take eskies full of fish back to the mainland. More and more are practicing catch and release, only keeping what they can eat while on the island. Make sure you are familiar with current bag limits and minimum sizes for different species, as well as any other restrictions (eg no fishing zones during the tailor spawning season) before you arrive on the island.

Leftover bait and fish remains should be buried just below high tide mark, covered with at least 50 cm of sand. Fish cleaning is prohibited in all campsites and campgrounds. Instead, use the fish cleaning facilities provided on the beach just north of the Pinnacles and Dundubara.

Since all freshwater fish are protected on the island, all freshwater creeks lakes and other wetlands (including the inland lakes) are closed to fishing. It is not possible to collect bait from these areas either.

## Environmental care

There are lots of things that you and your companions can do to minimise your impact and help protect the island's environment. With hundreds of thousands

*Sunset, Moon Point*

13

# Introduction

of people visiting the island each year, there is the potential for some people to become complacent. In doing the right thing, you will not only reduce your own individual 'footprint' but will be setting a good example to others.

If you've come to Fraser Island for the driving, camping or fishing opportunities, you're not alone. These activities, however, can have a serious impact on the environment. People who do the wrong thing not only spoil it for themselves and others, but call the future of these rights into question. Every driver, camper and angler therefore needs to be environmentally responsible. Most of the time, this just involves common sense. See pages 34 (driving), 8-10 (camping) and 12-13 (fishing) for tips on how to go about this.

Swimmers and campers can help to protect the inland lakes by ensuring chemicals such as sunscreen, detergents, shampoos, toothpaste, urine and insect repellents don't contaminate the water. Because what goes into the lakes stays in the lakes, or washes up on their beautiful shores, we all have a responsibility to look after these fragile places.

Smokers need to be conscious of how their habit can affect the island's fragile environment (as well as its visitors). Cigarette butts accumulate around the edges of the more popular lakes. If you're a smoker, why not collect your butts in an empty film canister like many of the backpacker groups are now doing? Cigarette bins are also provided at many of the more popular locations.

## Bushwalking

Bushwalking offers a pleasant way to take in the sights and sounds of the island, as well as escape the noise of other people and revving vehicles. Another attraction is the island's ever-changing vegetation which in turn supports an amazing variety

# Basic information

of bird and insect life. These features can be difficult to appreciate from a vehicle.

Bush walks range from short easy walks (eg Central Station to Wanggoolba Creek) to extended hikes that form part of the 90km Fraser Island Great Walk.

Information on the many options available as part of the Fraser Island Great Walk is available in the *Fraser Island Great Walk great info brochure*, which is available by calling the Naturally Queensland Information Centre on (07) 3227 8186 and online at www.epa.qld.gov.au/parks_and_forests/great_walks/fraser_island/. The brochure is also available from QPWS offices that issue camping permits (see p17) or can be downloaded at www.epa.qld.gov.au/publications. Some additional walks are listed at www.epa.qld.gov.au/projects/park/index.cgi?parkid=1.

The Fraser Island Great Walk includes numerous short walks (seventeen at the time of writing), for example:

Central Station to Wanggoolba Creek. 0.9km return (0.5 – 1 hour). Wheelchair accessible.

Central Station to Basin Lake. 5.6km return (2 – 2.5 hours).

Lake Wabby carpark to Lake Wabby. 3.1km return (1 – 1.5 hours).

Lake Wabby carpark to Eastern Beach. 7.3km return (2.5 – 3.5 hours).

Valley of the Giants to giant satinay. 7.3km return (2.5 – 3.5 hours).

Because the longer walks are only undertaken by a small number of people, they can make you and your companions feel as if you're the only ones on the island. However, these longer options may be more suited to experienced bushwalkers. While the terrain is not difficult, some tracks can be quite remote, exposed to harsh weather, and a little confusing to navigate at times. The section of the

15

# Introduction

*Inskip Point*

Great Walk north of Lake Wabby is described by QPWS as remote and suitable for experienced bushwalkers only.

Walking on sand can also be a challenge. While a bed of leaf litter covers many of the inland walking tracks making the going easier, other sections include exposed sand which can be quite soft in parts. Hikers need to be relatively fit, well-prepared and well-equipped. Take plenty of water, sun protection (eg hat, sunscreen), be aware of dingoes (see p 45) and snakes, advise a reliable person of your plans and let them know when you have returned. QPWS advises boiling or treating water from the lakes and creeks before drinking.

Make sure you obtain a copy of the brochure before you start planning your walk. As well as describing the different sections of the Great Walk, the brochure covers issues such as what to bring, what to do before you go and contingency plans. A Fraser Island Great Walk topographic map can also be purchased and is strongly recommended for hikers planning to stay one or more nights. This features detailed information about trip planning and preparation, safety, caring for the environment and track notes.

There are several camping options for those undertaking the longer walks. As well as beach camping and private camping options, a series of 'Great Walk camps' are provided at Jabiru Swamp, Lake Boomanjin, Lake Benaroon, Central Station, Lake McKenzie, Lake Wabby, Valley of the Giants and Lake Garawongera. These offer toilet facilities, seating, fresh water and small tent sites. Details of what is available at each site, including information on the water source and whether treatment is required, are provided on the website (www.epa.qld.gov.au/parks_and_forests/great_walks/fraser_island/). Because upgrades or changes may occur, it is worth obtaining up-to-date information prior to travel. Walkers are required to travel in groups of no more than eight people.

Before setting off, you will need to book your Great Walk campsites in advance as well as obtain a permit. Bookings can be made by visiting www.qld.gov.au/camping or phoning 131304.

# Basic information

## Permits

If you're making your own way over to the island, make sure you collect the Fraser Island Information pack prepared by QPWS before you arrive. This contains maps and important safety information.

Tourists bringing their own vehicles to the island are required to obtain a Vehicle Service Permit before arriving on the island. (You will receive your Fraser Island Information pack when you are issued this permit). The permit should be displayed on your vehicle's windscreen.

Campers planning to camp anywhere within the national park, whether free-range or in serviced campgrounds, require a Camping Tag (this does not apply to the three private campgrounds). The tag must be clearly displayed on your tent or camping structure. Hikers planning to stay overnight in any of the Great Walk campsites need to obtain a separate camping permit as well as book their sites in advance (see opposite page).

Vehicle Service Permits and Camping Tags are issued from the following offices and shops (note list not likely to be exhaustive and details may have changed).

**Naturally Queensland Information Centre**. Open 8.30am-5.00pm Mon-Fri. 160 Ann St, Brisbane. (07) 3227 8185

**QPWS Rainbow Beach office**. Open 7am-4pm daily. Rainbow Beach Rd, Rainbow Beach. (07) 5486 3160

**QPWS Maryborough office**. Open 8.30am-5pm Mon-Fri. Cnr Alice & Lennox Sts, Maryborough. (07) 4121 1800

**QPWS Bundaberg office**. Open 8am-5pm Mon-Fri. 46 Quay Street, Bundaberg. (07) 4131 1600

**QPWS Great Sandy Info Centre**. Open 7am-4pm daily. 240 Moorindil Street, Tewantin. (07) 5449 7792

**River Heads Permit Kiosk**. Open 6.15am-11.15am & 2pm-2.30pm daily. Barge landing carpark, River Heads

**Marina Kiosk**. Open 6am-6pm daily. Buccaneer Avenue, Urangan (Boat Harbour). (07) 4128 9800

Alternatively, both permits can be ordered online at www.epa.qld.gov.au (click on 'Camping bookings') or by calling 131304. The revenue raised from permits goes back into managing recreation on the island and providing facilities for visitors.

At the time of writing, a Vehicle Service Permit for a single vehicle costs $32.60 for any period up to one month, and $163.20 for one year. Camping fees are $4.00 per person per night or $16.00 per family per night (a family group being up to two adults and accompanying children under 18). Children under 5 are free.

Note that some visitors will require special permits (eg for professional photography / filming, guide dogs). Enquiries should be directed to the Maryborough office:

**QPWS Maryborough office**. PO Box 101, Maryborough Qld 4650. Open 8.30am-5pm Mon-Fri. Cnr Alice & Lennox Sts, Maryborough. (07) 4121 1800

## Ferries & barges

Fraser Island can be accessed in a number of ways including by aircraft, vehicle barge, passenger ferry or private boat.

If you're bringing a vehicle over, you'll need to book a place on one of the vehicle barges. (These are also available to foot passengers). For the barges that land on the beach, many people deflate their tyres just before leaving the mainland and engage 4WD before driving off the barge.

The Kingfisher Fast Cat passenger ferry transfers passengers from Urangan Boat Harbour (Hervey Bay) to Kingfisher Bay. Departs Urangan 8.45am, 12.00 noon,

# Introduction

4.00pm, 6.30pm - 7.00pm on Fri/Sat, 10.00pm. Departs Kingfisher Bay Resort 7.40am, 10.30am, 2.00pm, 5.00pm. Phone: 1800 072 555 or 07 4120 3333. Web: www.kingfisherbay.com. Cost: $44 return (adult), $22 return (child). Bookings are required.

Vehicle barges operate from:

- Inskip Point to Hook Point
- River Heads to Kingfisher Bay
- River Heads to Wanggoolba Creek
- Urangan to Moon Point

The following costs are for one vehicle, its driver and up to three passengers return. Enquire about other options (eg motorcycle, trailer, extra passengers, walk on passengers, one way only). Since prices and operating times may change, we suggest you check these with the barge company or your travel agent prior to travel.

**Inskip Point to Hook Point**

The Rainbow Venture and Manta Ray Barges operate continuously between Inskip Point and Hook Point between 6.00 / 6.30am and 5.30pm daily, with extended hours in peak holiday periods and on demand. The crossing takes about 15 minutes. Bookings are not normally required.

- Manta Ray

Locally owned and operated. Cost: $65. Phone Kosta: 0418 872 599. After Hours (07) 5486 8600. Web: www.greenbarge.com/.

- Rainbow Venture / Fraser Explorer

Cost: $65. Phone: (07) 5486 3154. Web: www.fraser-is.com/barge.htm.

**River Heads to Kingfisher Bay**

- Kingfisher Barges

Departs River Heads 7.15am, 11.00am, 2.30pm. Depart Kingfisher Bay 8.30am, 1.30pm, 4.00pm. The crossing takes about 45 minutes. Phone: 1800 072 555 or 07 4120 3333. Web: www.kingfisherbay.com. Cost: $115. Bookings are required.

**River Heads to Wanggoolba Creek**

- Fraser Venture

Departs River Heads 9.00am, 10.15am, 3.30pm. (Saturday extra sailing at 7.00am). Depart Wanggoolba Creek 9.30am, 2.30pm, 4.00pm. (Saturdays extra sailing at 7.30am). The crossing takes about 30 minutes. Cost: $115. Phone: 1800 249 122 or (07) 4125 4444. Web: www.fraser-is.com/barge.htm. Bookings are required.

**Urangan to Moon Point**

- Fraser Dawn

Departs Urangan 8.30am, 3.30pm. Depart Moon Point 9.30am, 4.30pm (1/4 to 31/8), 5.00pm (1/9 to 31/3). The crossing takes about 60 minutes. Cost: $115. Phone: 1800 249 122 or (07) 4125 4444. Web: www.fraser-is.com/barge.htm. Bookings are required.

## Fuel & supplies

Fuel and supplies, including camping gas, are available from the main settlements on the island - Orchid Beach, Kingfisher Bay, Cathedral Beach Resort, Eurong and Happy Valley. Fuel is considerably more expensive on the island so many drivers fuel up before leaving the mainland. A few even take jerry cans. When driving on Fraser island, you tend to go through considerably more fuel than you would normally. Always keep and eye on your fuel gauge and stop for a top up rather than risk the huge inconvenience and potential danger associated with running out.

Most basic supplies - ranging from bread and milk to fishing tackle and ice - can be bought on the island. Again, prices tend to be significantly higher than on the mainland. Between them, the island's main

# Basic information

settlements offer a range of takeaway, cafe and restaurant dining options as well as plenty of souvenirs.

## Phones & internet

At the time of writing, mobile phone coverage is poor or nonexistent across much of the island. You are unlikely to get a signal in the centre, along the east coast and northern parts of the island. Reasonable coverage, however, can be achieved from places like Kingfisher Bay along the west coast and the top of some dunes. Public pay phones are provided at the main settlements on the island - Orchid Beach, Kingfisher Bay, Cathedral Beach Resort, Eurong and Happy Valley - as well as the ranger stations (Central Station, Eurong, Dundubara and Waddy Point). You'll also find them at Indian Head and Yidney Rocks. If you can't bear to be offline, coin-operated internet access is available at Kingfisher Bay, Eurong, Happy Valley and possibly elsewhere.

*Drinking water, Eurong*

## What to take

Wherever you go on Fraser Island, if you are not part of an organised tour, there are a few general things you should carry with you. These include a first aid kit (and ideally a knowledge of basic first aid), cigarette butt containers if you smoke, a generous supply of drinking water, and a number of bags and containers (preferably strong lockable containers) for storing rubbish and food stuffs.

You should bring at least one good map, navigation aids, vehicle and camping permits, fire extinguisher, vehicle spare parts (eg spare tyre) and recovery gear (eg snatch strap and shackles). Communication tools such as mobile phones and UHF radios are also a good idea (although mobile phone coverage is variable - see above).

If camping, draw up a check list of what

*Indian Head*

# Introduction

you need before you go. While most basic supplies and a few camping accessories can be purchased on the island, it is usually better to avoid this extra expense and 'doubling up' on what you already have.

What you take will depend on the kind of camping you plan to do. It is advisable to bring a fuel stove if you are planning to cook as well as plenty of water for washing and drinking. It is also worth noting that it rains more on the island compared with the adjacent mainland so you may want to come prepared for wet weather (the region experiences a January-March wet season, but rain and storms can occur at any time of the year).

Other useful items include a torch, coins (for showers and pay phones), spade for burying toilet waste if you're bush camping away from toilets, and small and large strong lockable containers for storing food stuffs, bait and other items. This helps to reduce the amount of rubbish such as plastic bags as well as protect against animal scavengers. (It is still advisable to secure food and rubbish in a dingo lockup facility provided or a secure location like the back of your vehicle).

Generators are not permitted in any of the QPWS campgrounds or elsewhere in the national park apart from beach campsites. If you do decide to take a generator, you'll need to be considerate of others and minimise use after 9pm.

A door mat can be very handy for keeping sand out of your tent, and large long tent pegs help to better anchor your shelter in the sand and safeguard against strong winds. (You should not tie your ropes to nearby trees as this can damage them). If you're staying in QPWS-managed campgrounds, you will need $1 coins to operate the hot showers.

QPWS provides information on what you should not bring to the island. This includes things like firearms, fireworks and chainsaws, dogs (and any other animals), and unclean firewood, soil, plants or other media that could support disease.

## Safety

In an emergency (medical, police, fire), dial 000 (landline) or 112 from a mobile phone. Evacuation by helicopter may be required. Although at the time of writing there are no hospitals or health clinics provided on the island, an ambulance service operates out of Happy Valley during Queensland school holidays (phone: 07 4127 9158). A police station is situated at Eurong (phone: 07 4127 9288).

The vast majority of visitors to Fraser Island have a safe, enjoyable holiday experience. But like many other places in Australia, there is plenty of room to come unstuck. Some of the very things that attract people to the island - such as its remoteness, wildlife, and opportunities for recreation - can create problems.

Some visitors can be unlucky, others may simply be ignorant of potential hazards. Unfortunately, a few choose to engage in risk-taking behaviour. While we can't hope to cover the scope of potentially threatening situations in this book, here are a few tips that may help:

- drive safely, don't show off or hurry
- always have a generous supply of drinking water with you
- know about some of the animals to be wary of, especially dingoes (see p 45)
- take a first aid kit and be familiar with basic first aid
- take navigation aids and bear in mind that at times, tracks may be impassable or closed off by park authorities (always allow plenty of time)
- bring any personal medications - there are no chemists

# Basic information

- stick to walking tracks and stay with your group
- let a responsible person know where you are going and when you expect to be back, particularly if you are planning a hike or an extended trip
- don't sunbathe or allow children to play along the eastern beaches or anywhere where there is traffic
- don't swim along the eastern beaches because of rips, sharks and the lack of life guards
- guard against theft - keep your vehicle and belongings secure

## Ranger stations

Local QPWS staff can provide further information on the island:

Central Station Ranger Base
Phone: (07) 4127 9191

Dundubara Information Centre
Phone: (07) 4127 9138

Eurong Information Centre
Phone: (07) 4127 9128

Waddy Point Ranger Base
Phone: (07) 4127 9190

# Introduction

*A typical track in the island's west*

Fraser Island is widely regarded as one of the world's top island destinations. The old saying "half the fun is in getting there" rings true for most visitors.

As well as offering an abundance of spectacular natural attractions, the island has become a drawcard for offroad enthusiasts lured by the seemingly limitless opportunities for four-wheel driving.

Virtually all of the island's so-called roads are little more than sandy tracks and beaches and for many visitors, this is part of the appeal. If you're planning on driving, you will definitely need a four-wheel drive vehicle. Reasonable ground clearance is also needed for most of the inland tracks, particularly if track conditions have deteriorated due to very dry or wet weather or heavy vehicle use. Good clearance also helps with tricky sections or unexpected hazards along the beach.

Both the novice and experienced four-wheel driver can experience a variety of enjoyable offroad driving, from relatively easy-going tracks to trickier sections involving soft sand, tree roots, creeks and rocky outcrops. Track conditions on Fraser Island are constantly changing and can be difficult to predict. Be prepared for

*Pulling into a passing bay*

# Driving on Fraser Island

a wide range of possible driving scenarios and make sure you read the latest beach and track report issued by QPWS.

Circumstances can vary greatly depending on a range of factors including the weather, tides and amount of traffic. Because some of these changes can occur rapidly - in a matter of hours, even minutes - it always pays to be cautious and alert, even when tackling familiar ground. Night driving is best avoided and on the inland sand tracks, it is advisable to always have four-wheel drive engaged.

A general overview of sand driving is provided in the QPWS safety guide *Driving on Sand* which can be downloaded at www.epa.qld.gov.au/publications.

## Inland tracks

A large network of tracks crisscross the island. Even the simple act of crossing from one side to the other can take you through a range of track conditions.

The inland tracks vary from sections that are straight and flat through to those that follow the rise and fall of the dunes, and take in long sweeping corners. Wherever you are planning to go, make sure you give yourself plenty of time. The tracks all tend to be bumpy and slow-going. Any smooth sections are soon broken up with bumpy patches of tree roots that require you to slow right down. In dry conditions, sand can often build up into a mound running down the centre of the tracks, requiring reasonable vehicle clearance. For these reasons, the inland tracks are generally not suitable for caravans, or trailers or vehicles with low clearance.

The vast majority of inland tracks take two-way traffic, despite the fact that they are a single lane width.  At the time of writing, a speed limit of 35 km per hour applies on all inland tracks, except where stated otherwise. Driving conditions, however, often dictate much slower speeds

*Take extra care around blind corners*

(often only 15 km/hr will be possible). There are lots of blind corners for example which present the potential for head-on collisions and should be approached cautiously. As well as blind corners, there are many other occasions where the driver's line of view is obscured or there is some other reason to slow down.

Numerous passing bays (small clearings) are located along the sides of the tracks to allow vehicles to pass or overtake one another. You should expect to meet oncoming traffic at some point. Above all, common sense and courtesy are the key. If, for instance, you encounter a larger vehicle such as a tour bus or service truck, you should give way to them, simply because it is much easier to manoeuvre a smaller vehicle than a larger one.

Similarly, if you come across a convoy of vehicles or vehicles towing trailers, chances are it will much easier and safer for you to give way to them. In situa-

23

# Introduction

tions where circumstances appear equal for both parties, the vehicle nearest to a passing bay generally gives way. The passing vehicle should wait until the other is entirely within the passing bay. It is up to the drivers involved to decide on the best course of action at the time. It is not unusual for two oncoming vehicles to pull off the track simultaneously, both drivers wishing to give way. Drivers travelling uphill generally give way to drivers travelling downhill, especially in soft conditions.

Some passing bays can be very soft and may have significant ridges of sand to drive over when pulling off the track. If you're heading uphill, these may be very difficult to drive forward onto. The vehicle's front wheels slip more easily when faced with very soft sand on an incline. Reversing into the bay may be an easier option. This gives the front steering wheels firmer sand to turn on.

## Beach access points

Some of the softest sand you are likely to encounter is situated at beach access points (foredunes). It will be necessary to drive these sections when accessing the inland tracks from the beach and vice versa but also when driving around some of the bypass routes such as Indian Head. Beach access points are where the inexperienced driver is perhaps most likely to become bogged. Drivers should therefore be prepared by using appropriate tyre pressures and driving technique. Four-wheel drive should be engaged.

These soft access tracks and bypasses require a little momentum to get through. In manual vehicles, it is advisable not to change gears in very soft sand as this will cause your vehicle to slow rapidly, thereby losing momentum. The correct choice of gear to begin with depends on your vehicle and the terrain. In soft conditions, second gear high range is usually sufficient in our Land Rover. With experience, you'll gain a better understanding of which gears to use in different conditions.

Anxious to avoid a bogging, some drivers use far more speed than necessary when driving through these soft sections. This is a reckless practice, particularly if one or more bogged vehicles are in the vicinity. As well as the unpredictability of the speeding car's wheel placement, the drivers and passengers of the bogged vehicles are often distracted. Some may be lying down digging and virtually impossible to see. It is far better to wait for the bogged vehicles to free themselves, or offer assistance, than to risk lives.

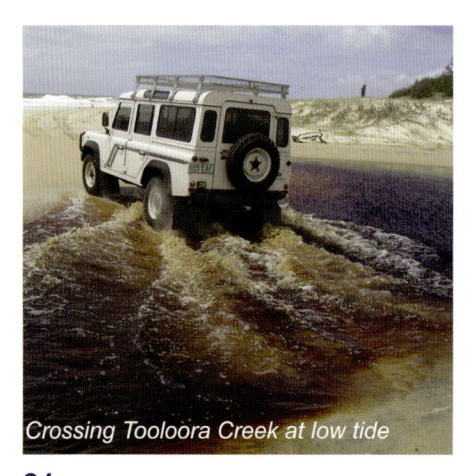

*Crossing Tooloora Creek at low tide*

*Look out for steep banks & other hazards*

# Driving on Fraser Island

## Beach driving

Beach driving is very different to driving on the inland tracks. On the beach, you will find yourself in a big open space, often with many cars traveling inline at quite high speeds. This can be an unexpected, sometimes daunting experience for first-time visitors.

On beaches that are safe to drive on, beach driving is best attempted at low tide. In general, these beaches are often inaccessible for at least two hours either side of high tide. Providing storm conditions are not forcing the water high up onto the beach, low tide usually produces a considerable width of beach to drive on. In these conditions, most drivers prefer to drive on the firmest section of beach (between the waterline and high tide mark), with many vehicles traveling up to the beach speed limit (80 km/hr at the time of writing). Know your tide times before you attempt to drive on the beach. Driving at low tide minimises your chances of having to drive through salt water or getting stuck somewhere because of a rising tide. Driving through salt water should be avoided.

Rules for driving along the beach are the same as those for driving on normal roads. This includes keeping left and indicating where applicable. It is not uncommon, however, for vehicles to pass on the opposite side as some drivers will either persist in driving on the firm sand nearer the ocean or refuse to change course in their effort to avoid the salt water. When encountering a situation like this, many drivers use their indicators to let oncoming vehicles know their intentions. A driver will use their left indicator, for example, to signal their intention to pass on the left-hand side of the oncoming vehicle.

Beach driving can look deceptively easy, with its firm sand and wide-open spaces, but this is often far from the truth. At times, some of the easiest driving can be had on the beach, but complacency and

*Be prepared for soft sand when driving on and off the beach, like here at Eurong*

25

# Introduction

the unexpected can spell danger for unwary drivers. Accidents occur every year and sadly, some have been fatal.

When driving on the beach, you should make your intentions clear and try to be predictable. Aim to avoid any sudden changes in direction. Slow down and drive cautiously when you see pedestrians or parked cars (eg around the Maheno Wreck and Eli Creek). Oftentimes, people will not hear the approaching traffic over the sound of the ocean. Use your commonsense.  Park your vehicle at an angle to the water, well away from traffic.

You should constantly be on the lookout for obstacles like mounds, ridges, creek dropoffs, rocks and logs or other debris. Even the smallest of these have the potential to destabilise your vehicle at high speed. Changing tides can leave seaweed, depressions and steep sandbanks that were not there before so always be on the lookout for the unexpected.

Unless you are following another car, such obstacles can be difficult - sometimes even impossible - to spot when travelling at high speed. Shaped by the wind and water, beach ridges, creeks and mounds are continually changing. Generally, these obstacles tend to be more abundant and severe higher up the beach which is why many drivers prefer to drive nearer the sea where it is comparatively flat. If you do choose to drive higher up,

it is advisable that you drive at a much slower speed, even when conditions appear flat.

Seat belts should be worn at all times and passengers should never be carried outside the cabin of your vehicle. Whilst these basic safety precautions and legal requirements should go without saying, it is disturbing how many drivers and passengers treat Fraser Island as an exception. All traffic rules apply on the beach and inland tracks and police issue tickets to offenders as part of regular patrols.

Be aware that planes take off and land on the beach between Happy Valley and Dundubara. Be alert and keep well clear, allowing them access to the firm section of sand.

## Creek crossings

Creeks present another potential hazard and should always be approached with caution. You will encounter numerous creeks along the length of Fraser Island's beaches. Although most of them are quite small and shallow, at times they are capable of cutting steep banks (washouts) that can be extremely hazardous, especially if driven into at high speed.

Tidal changes and storm conditions can alter the position of creeks out across the beach, changing an area you might have passed through only recently. When crossing a large creek like Eli Creek, aim for low tide and if possible, cross at its shallowest point (nearest the sea). Here, the increased width reduces the depth and velocity of the water as well as the steepness of the banks. Use a low gear and a little momentum and avoid changing gears midstream as this will cause your vehicle to temporarily lose momentum. You should always try to avoid coming to a complete stop in any creek bed.

If you have any doubts about  a creek

*Driving over Ngkala rocks at low tide*

# Driving on Fraser Island

*The beach is constantly changing - this creek dropoff formed overnight*

crossing, it is a good idea to get out of your vehicle and walk through first. When satisfied that the bottom is sufficiently firm and the crossing can be made safely, select the best route through by walking it. If you are still in doubt, it is probably best not to cross. Many of the creeks become impassable at high tide or in stormy conditions. Be aware that waves may surge up the creek, dramatically increasing the depth (not to mention saltiness!) within a matter of seconds. If conditions are unfavourable, you may need to wait it out or find an alternative route. While this might alter your plans, it's far better than losing your vehicle to the sea.

## Rocky outcrops

One of the more obvious hazards on the beach are the exposed outcrops of coffee rock; an organic-based material that can vary greatly in hardness. Although the amount of coffee rock visible on the beach is determined by tidal and sand movement, large, permanently-exposed areas of coffee rock are a regular feature of the island's beaches.

Large outcrops such as those at Poyungan and Yidney rocks often block what might otherwise be a continuous run along the beach. Most drivers take the inland bypass routes provided, however it is not uncommon to see drivers familiar with conditions driving on the beach around these rocks when the tide is sufficiently low. If you decide to do this, exercise caution. There have been numerous cases of vehicles becoming stuck and subsequently swamped by the incoming tide.

Before deciding to negotiate these rocky outcrops, check that the tide is sufficiently low by stopping and taking a moment to

# Introduction

*Coffee rock and seaweed along the western beach, near Woralie Creek*

study the waves. If conditions are favourable, plan the route that you wish to take and make sure it is clear. Busier times provide an opportunity for less experienced drivers to watch others and this may assist in choosing the best route. A few drivers decide to drive through shallow seawater to negotiate these rocks, rather than take the bypasses. This practice is not recommended and can result in expensive fines for hire car drivers.

Ngkala rocks, located to the north of Waddy Point, should only be negotiated by experienced four-wheel drivers with high clearance vehicles. These sections involve driving over rocky outcrops and using bypasses with often very soft sand. Many visitors to the island's north decide to travel as part of a convoy. It can be nice to know that an additional vehicle is on hand if assistance is required.

## Western beach

While vehicles are permitted along sections of the western beach between Moon Point and Wathumba Creek, conditions can be hazardous. The sand on the western side of the island can be very soft and unpredictable with buried seaweed.

There are also some large creeks that may be difficult or impossible to cross. It is important to inspect any water crossing on foot first to check the depth of the water and stability of the creek bed. If you intend to drive along any sections of the western beach you should be suitably experienced in sand driving and have the necessary recovery gear available. Ideally, it is best to travel with at least one other vehicle to aid recovery if needed.

## Tyre pressure

Reducing your vehicle's tyre pressure

# Driving on Fraser Island

*Reducing tyre pressure helps your vehicle to 'float' over the sand*

enables the walls of your tyres to flex, thereby increasing your tyres' 'footprints'. This creates better flotation in the soft sand since the surface area of tyre in contact with the ground is increased.

There are no hard-and-fast rules when it comes to tyre pressure and opinions tend to vary. This reflects the fact that there are so many different vehicles with different types of tyres. But as a general rule for soft, sandy conditions, air pressure should be reduced to around 20-25 psi. The best tyre pressure to use in a certain situation will depend on many factors such as track conditions, vehicle and tyre type (keep within manufacturer specifications), as well as the weight of your vehicle's load.

If conditions are very soft or if your vehicle is badly bogged, reducing tyre pressure further to around 15-18 psi

# Introduction

can be enormously helpful. In extreme circumstances, even lower pressure can be used however the downside is that this greatly increases the risk of separating the tyre from the rim, especially when sharply changing your steering direction. Whenever dropping tyre pressure below 20-25 psi, it is advisable to avoid driving at speed as the tyre will heat up and possibly fail. Tyres should be reinflated as soon as possible. Ideally, you should carry a tyre gauge and air pump. Reading tyre pressure is most accurately done when your tyres are cold.

## Driving technique

On the island's inland sand tracks, it is best to engage four-wheel drive as this minimises damage to the tracks and makes the going easier. Generally speaking, high range will be suitable most of the time. Remember that in soft conditions, the key to problem-free sand driving is reduced tyre pressure and sufficient momentum - this often means using higher engine revs than sealed road driving. Ideally, you want to keep up momentum without working the engine too hard.

Things like vehicle and engine type, the load you are carrying and prevailing conditions will dictate which gear to use. If in doubt, try high range first and change to low (if applicable to your vehicle) if the going gets tough.

During prolonged dry spells, track conditions can deteriorate significantly, with the deeper sand (usually moist) drying out as well. When this happens, it can be easy to lose traction, especially when driving uphill or attempting to move off from a stationary position. In such conditions, correct gear selection and momentum, combined with reduced tyre pressure, is very important.

Light rainfall and moist conditions usually improve driving conditions on the island.

The wet sand is more readily compacted and this results in a much firmer driving surface. However, the water runoff associated with very heavy rainfall can cause tracks to erode quickly. Wheel ruts may deepen and large puddles of water may form on the track. Although losing traction will be less of a concern, you will need to take care when passing through puddles where the depth and terrain beneath are unknown. You may wish to test the water first by walking through. If you decide to proceed, drive at a sensible speed. Large water splashes may cause petrol-engine vehicles to stall.

Since ocean beach driving is usually on firm packed sand, high range is usually best. The higher revs sometimes needed on the inland tracks are generally not required on the firm sections of beach.

Depending on the mode of 4WD, transmission wind-up can be an issue for certain types of vehicles on hard surfaces such as bitumen or hard-packed sand. Drivers should refer to their vehicle operating manual for information. Don't forget to re-engage 4WD again when you encounter soft sand or you may become bogged.

Another important thing to remember is to always keep your thumbs on the outside of the steering wheel. If not, you risk injuring your thumbs if the steering wheel suddenly 'pulls' to one side.

## Recovery techniques

Getting stuck in the sand, or 'bogged' as it is more often called, is a common occurrence on the island. Don't panic if you do come to a complete halt in the soft sand. Chances are, you will be able to do something about it.

It's best, of course, to avoid getting bogged in the first place. You can minimise your chances of getting stuck by driving with four-wheel drive engaged on

# Driving on Fraser Island

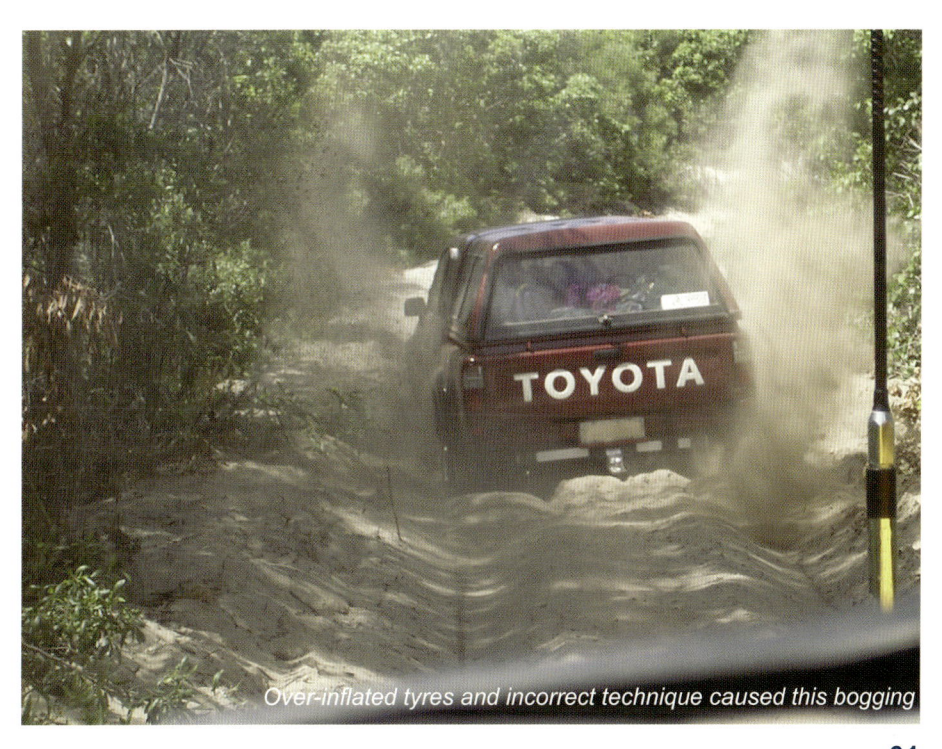

*Police conduct breathtesting to pick up drink drivers*

*Over-inflated tyres and incorrect technique caused this bogging*

31

# Introduction

the inland tracks, even when conditions may not seem to warrant it. If your vehicle is already in four-wheel drive, there is less chance of getting caught out if conditions suddenly change, plus you can save fuel and minimise damage to the tracks.

As mentioned, lowered tyre pressure and maintained momentum can make all the difference in sand. In vehicles with a manual gearbox, correct gear selection can also be crucial. A common mistake made by inexperienced sand drivers is to change gear in very soft sand. The moment they engage the clutch, drive is temporarily lost and the vehicle slows rapidly. The correct gear to begin with depends on the terrain, the type of vehicle and the load you are carrying. Second gear / high range is often sufficient for soft sections of the inland tracks. Experience will help dictate what will be required under particular conditions.

In the event of encountering difficult terrain and the vehicle beginning to slow, it is important not to panic. If possible, let the vehicle come to a halt on its own, stalling it if necessary. Once forward movement has ceased, it will generally be futile to try and proceed by continuing to turn the vehicle's wheels in the same direction that you were travelling. In doing this, you risk worsening the situation by 'digging' your vehicle further into what is potentially bottomless soft sand.

Instead, gently reverse as the sand you have already driven on will be slightly compacted. If this fails, try 'rocking' the vehicle by alternatively selecting forward and reverse gears and moving the vehicle a little at a time (backwards and forwards) until you have sufficiently compacted the sand in order to proceed. In most circumstances, this is all that will be required to get the vehicle moving again. If you only succeed in moving a short distance, the entire procedure may have to be repeated

*Many drivers bog their vehicles at the Indian Head bypass*

*Check the water's depth before you proceed (track to Moon Point after heavy rain)*

32

# Driving on Fraser Island

- perhaps several times - until you reach firmer ground.

In more severe circumstances, where the above efforts fail to extricate the vehicle, there are several options open to you. Some are possible to perform on your own. Others require the assistance of another vehicle. It will depend on the situation and what assistance and / or recovery gear is available.

A spade or a long-handed shovel should be carried in your vehicle. Digging may be necessary to remove built up sand from in front of the wheels and possibly from underneath the vehicle. For example, sand may have banked up in front of the axles. Some vehicles have been freed after being physically pushed by a small group of people. If this is attempted, those pushing need to be aware of sand sprayed up by the vehicle's spinning wheels.

A snatch strap and a pair of rated bow shackles are very handy pieces of equipment to carry. Although simple to use, they require the use of a second vehicle. Make yourself familiar with their correct use before you need them. Use the shackles to connect one end of the strap to the bogged vehicle and the other to the recovery vehicle. It is very important to utilise the vehicles' correct recovery points when doing this. (Some vehicles will have a hook for attaching the strap so the shackles will not be required). The two vehicles should be close enough to allow the strap to be quite loose, not taut.

A snatch strap works a bit like a rubber band. As the recovery vehicle is driven away, the strap stretches then contracts, snatching the bogged vehicle free. In general, the driver of the bogged vehicle should attempt to drive out at the same moment that the strap contracts, remembering to keep the wheels in a straight line. Bystanders should keep well clear in case the strap or shackles break and catapult towards them.

If bogged in very soft sand, lowering tyre pressure even further (eg to around 18-psi) further increases each tyre's 'footprint'. Drivers are often surprised at what a difference this can make. It is possible to run an even lower tyre pressure if circumstances warrant it (see p28-29).

In severe circumstances, where there is no other alternative, material may need to be placed on the sand to aid traction. In some situations, this requires being inventive. Branches and other plant material are commonly used but remember that obtaining them recklessly will cause harm to the natural environment. Fines can be imposed for any damage to the natural surroundings. If you have no other option, it is best to use dead wood lying on the ground, making sure there are no sharp edges that could damage your tyres. Once the material has been used,

*Using a snatch strap to recover a bogged vehicle*

# Introduction

it should be put back exactly where it was found. Even dead wood provides habitat and helps to stabilise the sand.

National park rangers often spread wood chips over problematic sections of track, while wooden boards and rubber have been laid down on many of the steeper dune tracks to aid traction.

## Environmental driving

Fraser Island's natural qualities are its greatest asset. The freedom to be able to explore what is arguably one of the most beautiful offroad destinations in the world should never be taken for granted. At all times, drivers should treat the island with the care and respect it deserves. In doing this, they can help to ensure this privilege will be around for many years to come.

On an island teeming with wildlife, you are likely to come across an animal on the inland tracks at one time or another.

Dingoes and birds are commonly sighted but animals such as snakes, lizards and turtles may also enter your vehicle's path. If it is safe to do so, slow down - even stop - until the creature has passed. There is no need to handle animals, they will usually move of their own accord. Remember you are in their territory. Keep a safe distance from dingoes and snakes.

Make sure everything in your vehicle is secure. You don't want plastic bags, rubbish or other items to blow out of the vehicle and potentially cause harm to wildlife.

Although there is a sign alerting drivers to the problem, birds - particularly terns - are sometimes struck by cars traveling at high speed along Seventy-Five Mile Beach. When approaching birds on the beach, it is advisable to slow down and go around if safe to do so. Allow dingoes to pass.

Some birds, reptiles and other animals nest in the dunes and in the banks of creeks. For this and many other reasons, it is important that you avoid driving on dune, creek and beach vegetation, including grasses and small ground cover plants. At all times, keep to designated tracks and camping areas.

On the inland tracks, use existing passing bays when moving aside for oncoming vehicles and avoid driving over vegetation - even if the nearest passing bay is some distance away and requires one of you to reverse. If you're setting up camp, keep to pre-existing tracks and clearings.

You can minimise damage to the inland sand tracks by always having your vehicle's four-wheel drive engaged.

Never indulge in reckless behaviour such as attempting to drive up the face of a steep dune as this causes irreversible environmental damage. Driving on dunes is prohibited. Common sense and a respect for every inch of the island is the key to environmentally-responsible driving.

# Driving on Fraser Island

## Quick tips

This summary will help to reinforce some of the information provided in this introduction to driving on Fraser Island.

It is not intended to be a substitute for reading this section in full or for the more extensive research of sand driving we strongly recommend you undertake before setting off (beyond the scope of this book). At a minimum, you should read the QPWS safety guide *Driving on Sand*.

### Tyre pressure

There are many opinions on the subject of tyre pressure, mainly because there are so many different vehicles with different types of tyres. Typically, a pressure of around 20-25 psi works well. But if you find yourself in trouble, you can go lower (~18 psi). If you do, remember to take things easy and reinflate as soon as possible.

### Gears

Correct gear selection that allows you to keep up momentum without working the engine too hard is important. Things like vehicle and engine type, the load you are carrying, and prevailing conditions will dictate which gear to use. If in doubt, try high range first and change to low if the going gets tough. Engage 4WD on inland tracks.

### Speed

Drive according to the conditions at the time, always obeying speed limits. Stay continually alert, take extreme care when approaching blind corners on the inland tracks and be aware of the ever-changing conditions on the beach. Slow down when approaching creeks, people and parked cars and aim to avoid any sudden changes to your vehicle's direction.

### Tides

Know your tide times before driving on the beach. Driving at or near to low tide makes good sense and minimises your chances of having to drive through salt water or getting stuck somewhere because of a rising tide.

### Recovery gear

It is advisable to always carry a snatch-strap and rated shackles and make yourself familiar with their correct use. Also make sure you have a shovel or spade handy for digging yourself out of a bog.

### Recovery procedure

If you come to a stop don't panic. Continuing to spin the wheels will usually only dig your vehicle in further, making things worse. First, try to gently reverse out following your tyre tracks. You may need to 'rock' backwards and forwards. Dig away sand that has built up in front of your tyres. If this fails and a second vehicle is present, you might want to try 'snatching' the vehicle out. In very soft sand, consider lowering your tyre pressures further (if you haven't already). Keep yourself and passengers well clear of traffic and snatch attempts.

35

# Introduction

## Aboriginal history

For thousands of years, Fraser Island was inhabited by Aboriginal people known as the Butchulla. Evidence of their remarkable culture and tradition can be found in the many archaeological remains that still exist across the island today as well as the stories and memories kept alive by today's Butchulla descendants.

The Butchulla named the island after the beautiful 'white spirit' they called 'K'gari' which can also mean paradise. All up, six clans of the Butchulla tribe are believed to have occupied different areas of island. (Butchulla - meaning 'sea people'- is also spelled Butchalla, Butchella or Batjala.)

Not restricted to the island, the Butchulla people's mainland territory extended across the Great Sandy Strait to Bauple Mountain. At times, they would canoe across to the mainland using boats made from long strips of bark, tied and sealed with wax. Fire would be carried inside the boat on a bed of sand or seaweed and used for cooking freshly-caught fish.

The island population of approximately 400-600 people would dramatically increase during winter when seafood was abundant. While these numbers may seem low compared with the hundreds of thousands of tourists that now flock to Fraser Island each year, the island at that time in fact represented one of the most densely-occupied areas in Australia.

A healthy people, the Butchulla were nourished by a diet largely consisting of seafood. Fish, shellfish, crustaceans (crabs and yabbies), turtles and dugongs were among the items eaten. One of the most abundant resources, and a favoured food source, was eugarie (pipis or cockles). Middens containing thousands of shell remains serve as evidence and can be sighted today. Other archeological evidence of the Butchulla include artefact scatters, fish traps, scar trees and camp sites.

Plants provided an important source of carbohydrate and vitamins. Plant foods included particular nuts, roots, reed and grass seed, and fruits. Passed on from one generation to the next over thousands of years, Indigenous knowledge of how to collect and prepare these items was very detailed. The orange fruits of the pandanus palm, for example, needed special preparation to leach out the toxins. By modern standards, the Butchulla's diet was relatively bland. They are said to have valued sweet things such as the honey produced by small native bees and the nectar of certain flowering plants.

Using the natural materials available to them, the islanders constructed shelters from bark strips and stakes. While their clothing was limited to only headdresses and necklaces, in the colder months they would use fire and furs for warmth. Tools and utensils used for hunting and cooking were often made from rock. Since very little rock was available on the island, most of this is thought to have been sourced from Big Woody Island (situated between Fraser Island and Hervey Bay).

Totems were an important part of Butchulla cultural traditions. Each person was assigned a plant or animal totem. These tended to be non-abundant resources. Under this system, the person was not allowed to hunt or eat their totem on their own tribal lands, except during tribal wars and special occasions. They were also required to respect the totem of their mother, father, wife and her family.

The Butchulla social system – characterised by taboos, rituals, polygamy and male and female 'businesses' - has been described as harsh and complicated. Male initiation ceremonies, during which scars would be inflicted across the young man's chest and shoulders, were held in

# History

bora rings. Young girls were taught about the issues of womanhood and usually married off by the age of fourteen.

But the Butchulla's strong traditions and culture would eventually receive a powerful blow. In 1770, the famous explorer Captain Cook sighted a group of Aborigines on the island, naming 'Indian Head' after them. Captain Matthew Flinders (1799 and 1802) was among the first white men to have peaceful contact with the islanders. This is followed by accounts of escaped convicts coming into contact with and sometimes living with the Aborigines. At times, white people were regarded as the reincarnated spirits of dead family members.

In 1836, a ship named the Stirling Castle was wrecked just north of Fraser Island. The demise of the Butchulla people started soon after. Not long after the survivors came ashore, the ship's captain, James Fraser, was supposedly speared and later died while his wife Eliza Fraser (after whom the island was later named) was enslaved. Although Eliza managed to survive her ordeal and become famous for telling the story of her experiences at the hands of the Aborigines, many people say she exaggerated and over-embellished the details. This had the effect of creating little sympathy for the Aboriginal people who had probably saved her life.

In the years that followed, many Aborigines were killed in reprisal for the death of a few whites such as James Fraser and shepherds whose flocks were attacked. The dispossession of the local Butchulla people had begun. From the 1850s, police are said to have orchestrated numerous attacks on the Aboriginal population of the Wide Bay area, with many massacres taking place. In one large-scale massacre said to have taken place on Christmas Eve in 1851, police (amongst whom were non-local Aborigines) were dispatched for a raid on Fraser Island.

1870 saw the establishment of a mission

*When ripe, pandanus fruis were prepared in a special way to remove toxins*

# Introduction

for Aborigines at White Cliffs (Balarrgan) on the western side of the island. This was later closed down, however, when the area was earmarked for a quarantine station. European agricultural practices had also proven unsustainable because of the unsuitability of sand for crop growth. In 1897, government policy forced the few remaining Aborigines in Maryborough to again be sent over to the abandoned quarantine station at White Cliffs. The 'new' mission was under the control of Archibald Meston - a man who respected Aboriginal culture and is said to have spoken numerous Aboriginal dialects. For a short time, some people deemed the mission a 'success'.

This 'success', however, was short-lived as tourists from Maryborough and people with shipping and timber interests pressured the government to close the mission. As a result, the mission was relocated to a less than ideal site at Bogimbah Creek, further north.

Aborigines from all over Queensland were then sent to the mission on Fraser Island with no consideration of differences between tribes and dialects. Between 50 and 200 Aborigines are thought to have been housed at the Bogimbah mission at any one time. Conditions were poor and many people died from disease. Regarded a failure, the mission finally closed in 1904 to make way for a logging camp. Most of the remaining Aborigines were then sent to Yarrabah near Cairns and Cherbourg, 200 km northwest of Brisbane, signalling the end of traditional Aboriginal occupation of the island.

Despite the many challenges they have had to face and continue to face, today's descendents of the Butchulla people are endeavouring to rebuild their cultural and spiritual ties with Fraser Island through their oral history and special organisations and projects.

## European history

As far back as 1521, the Portuguese may have been the first Europeans to record Fraser Island. Captain Cook later caught sight of Indian Head in 1770 on his Voyage of Discovery and Matthew Flinders explored the coastline in 1799. Clay pipes unique to Dutch traders have been discovered in Aboriginal middens near Indian Head, suggesting the Dutch may have also passed before Cook.

The first permanent European settlement on the island came with the construction of the Sandy Cape Lighthouse in 1870; the state government's aim being to reduce the number of ships lost in the area.

Interests in forestry began back in 1842 with Andrew Petrie's discovery of stands of cypress pine on the west coast. In 1863, 'Yankee Jack Pigott' extracted kauri pine and in 1905, a tramline was constructed from Urang Creek to the Poyungan and Bogimbah forests. Shortly after, in 1908, the central part of the island was declared a Foresty Reserve. Logging continued on the island for the next 83 years.

Timbers taken included kauri pine, hoop pine, white beech and cypress pine, followed by tallowwood, blackbutt, satinay, and brushbox. Central Station was established as a forestry base camp in 1920. The forestry headquarters were moved to Ungowa in 1959.

During World War II, Fraser Island was the site of secret military training. These activities included Z Force training just south of Kingfisher Bay and at Lake McKenzie, and practice bombing raids on the wreck of the Maheno.

Although the conservation value of Fraser Island was being argued as early as the late 1800s, it wasn't until after the arrival of sand mining that protective measures were finally put in place.

# History

Fraser Island's dunes contain sought-after minerals such as zircon, rutile, ilmenite and silica.

Dilli Village was constructed as a sand mining base in 1974, named after the large American industrial company, Dillingham. Together with Murphyores, the mining consortium's aspirations sparked many years of public outcry against plans to extract mineral sand from thousands of hectares of pristine dunes. Leases covering 8665 hectares had been granted as far back as 1966 and more were being sought. By the end of 1971, Queensland Titanium Mines had begun a mining operation between North Spit and Eurong. Led by John Sinclair and the Fraser Island Defenders Organisation, with the enlisted support of the Australian Conservation Foundation, the public opposition was heard in the Queensland Supreme Court and then the High Court of Australia.

In 1976, the Australian Government banned the export of minerals from the island. The publicity associated with this issue drew attention to Fraser Island as a tourism destination.

## World Heritage listing

More recent conservation efforts focussed on logging, prompting a Queensland Government Commission of Inquiry into the Conservation, Management and Use of Fraser Island and the Great Sandy Region in 1990.

Thanks to the persistence of conservationists, in 1992 the island was World Heritage listed and today, nearly all of the island is protected as national park. In terms of its World Heritage values, the island provides an outstanding example of ongoing ecological and biological processes as well as superlative natural phenomena.

The region's many natural and cultural features relate to its Aboriginal heritage, geomorphology, hydrology, vegetation, wetlands, estuaries, marine wildlife, ecological processes, diverse habitats and wilderness quality. Different aspects of each of these are described throughout this book.

Given the incredible natural appeal of Fraser Island, it is not surprising that hundreds of thousands of tourists now visit the island each year. If this privilege is to continue, however, it is important that every visitor makes an effort to minimise their impact.

# Introduction

*Banksia cone*

# Flora & fauna

Whether fascinated by the natural sciences or not, most visitors are amazed by the diversity of plants and animals flourishing on Fraser Island.

On a typical day, an observant visitor can enjoy a range of wildlife experiences. They might wake to the sound of more than a dozen bird species, spot a dingo or two scavenging along the beach, watch large birds of prey circling overhead and catch sight of a sand monitor dashing up a tree. While swimming, they might spot an eel or small native freshwater fish and at night, marvel at aerobatic microbats catching insects, later falling asleep to the rings and chirps of tiny acid frogs.

Situated in one of Australia's most biodiverse regions, Fraser Island supports a diversity of vegetation and wetland types. These, in turn, support a myriad of animals and other life forms. Because of the area's geographic position and subtropical climate, the outer limits of many species' ranges overlap. In other words, the island marks the most northern limit for many southern species and vice versa. As well as encountering species new to science, scientists have discovered peculiar combinations of plants and animals that occur nowhere else in the world.

But in spite of this amazing diversity, the island's wildlife isn't always as obvious as some people might hope. Don't expect grasslands overflowing with herds of grazing kangaroos or spectacular nocturnal displays by acrobatic sugar gliders. Some people's perceptions of wildlife have perhaps been distorted by what they've seen in television documentaries, zoos and the cleared agricultural landscapes of the Australian mainland.

Fraser Island's wildlife doesn't 'perform' or appear on cue, but what it does do is surprise, captivate and continually delight. Many people, for instance, are surprised to learn that more than 350 bird species have been recorded for the island, easily qualifying Fraser as one of the world's premier bird-watching destinations. But that's just the beginning - 300 species of ants, more than 900 plants, 74 reptiles and at least 17 native frogs call this massive sand island home.

Although only a handful of species are mentioned here, those chosen will hopefully get you thinking beyond just dingoes and whales. From inconspicuous nutrient recyclers, to bizarre insects and giant earthworms, Fraser Island is home to so many life forms; many superbly adapted yet poorly understood.

Every now and then, visitors are treated to an impressive spectacle and while plenty of patience and quiet observation can help, usually it's just a case of being in the right place at the right time. Binoculars and field guides can be great to have on hand. Pack a torch for spotting wildlife at night, taking care not to shine the light directly into any animal's eyes.

If in doubt, always keep a safe distance. Never provoke or corner animals, damage their habitat or attempt to feed or interact with them. Likewise, common sense should dictate that you shouldn't sample possible bushtucker plants unless accompanied by a local expert. Nor should you remove plant or animal material from the island - whether dead or alive. As the Butchulla elders advise "Do not take what is not yours."

## Island of birds

Wetlands, rich expansive mudflats and a myriad of inland habitat types are just three of the reasons why Fraser Island is a bird watcher's paradise. With more than 350 visiting and resident bird species, the area is said to support more bird species than any other region in Australia.

To better appreciate this diversity, a field guide and pair of binoculars will come in

# Introduction

seen or heard throughout the day, early morning and mid to late afternoon is when they are likely to be most active. Patience and a bit of good luck are often the key.

Tens of thousands of migratory birds use the island as a vital resting place after making the long journey from their northern breeding grounds. Some come from as far as Siberia. An important summer stopover, the region's coastal wetlands are utilised by 18 of the 24 migratory wader species listed under the Japan-Australia Migratory Bird Agreement (JAMBA) and China-Australia Migratory Bird Agreement (CAMBA). Amongst these are the lesser golden plover, large sand plover and red necked stint. At least 20 different species of migratory wading birds roost in the region.

Bar-tailed godwits, Mongolian plovers, black-naped terns, Latham's snipes and grey tailed tattlers congregate along the island's shores. At least 22 different types of gulls and terns have been recorded. Eastern curlews are often seen at low tide from August to March. These large waders are recognised by their long curved bills and mournful cries. Whimbrels, which are similar in appearance but somewhat smaller, arrive in spring and sooty and pied oystercatchers (or pipi catchers) are commonly spotted foraging along the beaches. As their name suggests, oystercatchers are perfectly equipped to prise open tightly-closed bivalve molluscs using their specialised beaks. These locally nomadic, gull-like birds, which pair up for

*A plover prepares for flight*

handy but generally the trick to successful bird watching is to look and listen carefully. This can be done by sitting still and observing or by walking quietly, ideally through a range of habitat types. Alternatively, you might prefer to take to the water on a kayak to explore a mangrove-lined creek or one of the island's many lakes. Although a variety of birds can be

*Pipi catchers pair up for life*

# Flora & fauna

life, are often seen feasting on shellfish along the island's surf beaches.

Birds of prey, like goshawks, swamp harriers, sea eagles and osprey scan the land below for prey or carrion. Nearly 20 such species have been recorded in the area, including four species of falcon. But perhaps those most commonly seen are the brahminy and whistling kites. One of the best places to spot them is along Seventy-Five Mile Beach. Both kites display a slow, wheeling flight as they scan the dunes for potential meals.

Osprey are large, specialist fish catchers that often construct sizable nests in the forks of tall trees atop sand cliffs or rocky outcrops like Indian Head. A pair may use the same nest for many years. White-breasted sea eagles use their impressive two-metre wingspans to soar on the air currents, impressing lucky onlookers with their spectacular dives for fish.

The island's forests are incredibly rich in birdlife, offering vital habitat for many of the smaller species (too numerous to cover here). On the mainland, many of these birds are silently being impacted by habitat loss and invasive species. These forest birds exploit the island's enormous diversity of plant life and habitat types, particularly flowering heath and open woodlands where there is an abundance of insects, nectar and tree hollows. A variety of doves, cockatoos, parrots, thornbills, flycatchers, owls and woodswallows can be seen or heard along with several types of monarchs, flycatchers, butcherbirds, whistlers, warblers, cuckoo-shrikes, wrens, robins and shrike-thrushes.

Honeyeaters are well represented, with nearly a dozen different species having been recorded. Amongst the most commonly-sighted are the Lewin's, white-cheeked and scarlet honeyeaters. Eastern yellow robins flit through the forests' uninterrupted corridors of dense understorey while satin flycatchers are known to frequent resorts and campgrounds to feast on insects attracted by the lights.

Six species of kingfisher can be spotted in the forests and in the vicinity of freshwater lakes and creeks. These include the superbly coloured sacred and azure kingfishers, as well as Australia's famous laughing kookaburra. Another colourful bird, the rainbow bee-eater, nests in the banks of creeks and sand cliffs, while visiting flocks of rainbow lorikeets fill the forest with their loud, chaotic chatter.

*The kookaburra is the largest kingfisher*

Well-camouflaged hunters, like tawny frogmouths and nightjars, emerge under the cover of darkness to feed on small marsupials and rodents. During the day, bar-shouldered doves, willie wagtails and Torresian crows often forage in the campgrounds, while the distinctive whip-crack call of the eastern whipbird may be heard echoing through the rainforest.

Aquatic birds, although present, are

43

# Introduction

not as abundant as some visitors might expect given the island's large number of water bodies. Grebes, ducks (including the musk duck and wandering whistling-duck), and even brolgas are sometimes spotted, but the lack of nutrients in most of the lakes keeps numbers low. Ocean Lake, with its comparatively high nutrient load, is one of the few exceptions. This tannin-stained window lake, situated in the island's north, probably supports the highest population of breeding and nesting waterbirds on the island.

Fraser Island also offers vital habitat for several rare, threatened and otherwise-significant species, including the brush bronze wing, black breasted button quail and little tern. But perhaps the rarest is the ground parrot, a beautiful bright green, yellow and black coloured parrot that emits an unusual bell-like call. Ground parrots are restricted to heath and sedgeland habitat. If startled, they fly into dense vegetation then run.

## Dingo's domain

Fraser Island's dingoes fascinate visitors on a daily basis. No stranger to controversy, this wild dog population is thought to represent the purest strain of dingo in eastern Australia, its geographical isolation having limited opportunities for cross breeding with domestic dogs.

Scientists estimate dingoes first arrived on Australian shores from Asia about 5000 years ago. There is even recent genetic data to suggest today's dingoes trace back to a single pregnant bitch. Possibly descended from the Indian Wolf, dingoes are thought to have more in common with today's wolves than with domestic dogs. Dingoes howl, whereas most domestic dogs bark. Important for communication, a howl projected from the top of a dune might mean "come and join me" or "stay away" if directed at outsiders. Dingoes tend to have large, pointed ears

*Dingo-proof lockup units are provided at some locations*

and tan coloured bodies but are occasionally black, black and tan or all white.

Studies indicate that about 80 percent of the island's dingoes might be pure and that the purest occur in the island's north where dingoes have had less contact with humans. The northern dingoes have been observed to be more active at night, naturally more timid and wary of people and more territorial towards other dingoes.

Being social animals, dingoes prefer to live in packs. Each pack has its own territory within which to hunt, scavenge, take shelter, socialise and rear young. Not a lot is known about different dingo territories on Fraser Island, although dingoes are thought to spend quite a lot of time scent-marking along their borders. The idea is to keep other dingoes away, including bachelor groups and 'loners', with territories and mating rites defended aggressively against such threats. Each pack is governed by a dominant Alpha male and female. Just like wolves and domestic dogs, dingoes raise their tails, which are white-tipped, to signal dominance. This kind of behaviour is thought to be innate (inherited) whereas submissive displays of behaviour are mostly learned from other pack members.

Since there is only room in the pack for one Alpha male, it is not unusual for younger males to become outcasts or

# Flora & fauna

## Dingo safety

QPWS provides detailed information on how visitors and residents should and shouldn't behave around dingoes. This is regularly updated and comes with your camping and vehicle permit.

Set aside some time to read their information and any updates carefully, and make sure you and your companions talk about your first dingo encounter and how you will handle it before it happens. The brochure *Be dingo-safe!: Fraser Island World Heritage Area* can be downloaded from www.epa.qld.gov.au/publications.

For more peace of mind, camping groups with children often choose to camp in a camping area with a campground ranger. These areas are Central Station, Dundubara and Waddy Point. Fenced campgrounds are provided at Dilli Village, Lake Boomanjin, Central Station, and Waddy Point.

Always follow the directions given by QPWS signs, brochures and park rangers. Any incidences involving dingoes should be reported to a QPWS officer as soon as possible.

The *Be dingo-safe!: Fraser Island World Heritage Area* brochure, includes the following advice at the time of writing:

- Always stay close to your children, even small teenagers.
- Walk in small groups.
- Watch dingoes quietly from a distance; don't encourage or excite them.
- Look out for dingoes - stay calm and don't run.
- NEVER feed dingoes.
- Lock up your food stores and iceboxes.
- Pack away your food scraps.
- Keep fish and bait in sealed containers off the ground.
- Make your tent or house boring for dingoes - keep your belongings safe.
- Tell others how to be Dingo-Safe!

It's a good idea to contact QPWS or check their website for up-to-date advice prior to travel.

# Introduction

'loners' upon reaching maturity. These are frustrating times for a young male and he will spend a lot of his time trying to infiltrate groups and mate with females. Loners and bachelor groups are watched closely by rangers in case their bold behaviour becomes directed towards people.

Little is known about the relationship between Fraser Island's Aborigines - the Butchulla people - and dingoes. However, the association between dingoes and Indigenous people is believed to have often been a positive one throughout Australia. Fraser Island's dingoes may have been kept as 'living blankets'; others say their purpose was to guard rather than provide warmth or companionship.

Female dingoes generally reach maturity at two years of age, from which time they are able to give birth to one litter per year. (Domestic dogs mature much earlier and can breed up to two times per year.) Most puppies are born in the cooler months (June - August) and the mother dingo is very protective of her litter. It makes sense for a dingo to have fewer puppies if food or territory is more limited so litter size can range anywhere from one to ten puppies, depending on a range of factors

*Moving as a pack along the beach*

including how well nourished the mother is. It is thought that, on average, only one pup in the litter survives to maturity.

A puppy's life is fraught with danger. Like wolves, pack members are believed to help one another rear the offspring. But there is also a darker side to these helpful interactions. If more than one female from the same pack gives birth to a litter at the same time - and times are tough - the dominant female may kill off her packmate's puppies. Despite this seemingly gruesome act, the subordinate female then assists the Alpha female with her litter. The pups now have two mums, less competition, double the milk and a much better chance of survival. Weaning takes place some time between September and November and puppies must learn survival skills from older pack members like hunting, howling and finding water.

At this age, dingo pups are very inquisitive and appealing to humans. If they are to survive, however, it is essential people do not to interfere with this crucial learning process by trying to feed or interact with them. Throughout these months, it pays to keep well clear of dingoes. The same applies during the mating season (April to July) when young male dingoes trying to infiltrate packs and mate with females may be bold and agitated.

When it comes to food, the island's dingoes are adaptable and opportunistic, employing a combination of hunting

# Flora & fauna

*Native spinifex helps to stabilise the dunes*

and scavenging techniques. Night-time is the best time to hunt, since most of the island's mammals are nocturnal. Bandicoots, bush rats, bats, frogs, possums and gliders are all potential meals, though not very easy to catch. During the day, dingoes are often seen scouring the beaches for fish scraps, crabs, dead birds and whatever other leftovers they can find. They also eat seeds, vegetation, insects, lizards and turtle eggs.

Food offered by people can cause stomach infections because of the bacteria it contains. Hand feeding or failing to stow rubbish securely also encourages scavenging and dependency on humans as well as increased litter sizes and puppy deaths when there is not enough food to go around. As well as losing their natural hunting skills, over-fed dingoes become slow and unfit to hunt. When less food is available, dingoes are more likely to starve and die.

There is a terrible consequence of human feeding that cannot be emphasised enough. Dingoes that lose their natural fear of visitors are more likely to act unpredictably or aggressively.

In the past, authorities have destroyed dingoes considered to be aggressive and unafraid of humans that persist in populated areas. Such action was taken following the tragic death of a nine-year-old boy from an attack by dingoes in 2001. Tough penalties now apply for people who do the wrong thing. Remember it is an offence to feed a dingo, attract it using food or food waste, or disturb it in any way.

## Life in the dunes

Despite the harsh conditions, Fraser Island's sand dunes are literally teeming with plant and animal life. Because of the island's former connection with the mainland, the bulk of these species rely on pre-existing adaptations and in-built

47

# Introduction

*Above & right: Ghost crabs excavate burrows in the sand*

*Below: Goat's foot pioneers the dunes*

# Flora & fauna

resilience to not only cope but flourish in this dynamic sand environment.

The youngest dunes, situated along the eastern beaches, reveal the beginnings of plant succession, soil enrichment and dune development. Colonising plants like spinifex thrive in the moist, exposed conditions, sending out long runners that help to stabilise the sand. Their lateral growth may be encouraged by sand burial, so in areas where sand is actually building up, the grass responds with faster growth that in turn helps to trap the new sand. This paves the way for other pioneer species like goat's foot, pig face, flax lilies, banksias and sheoaks. Over time, more complex vegetation communities take over and, if succession is allowed to continue, a so-called climax community is achieved (closed forest or rainforest).

Fraser Island is one of the few places on earth that offers a time series 'snap shot' of succession in action. As you travel from east to west, dune age progressively increases and the type of vegetation matures accordingly. But for fascinated scientists, there was an unexpected turnaround. This process of succession begins to work in reverse - 'retrogressively'. The proof lies in the vegetation clothing the older dunes in the island's west. On this side of the island, thousands of years of leaching have pushed the nutrient-rich soil layer to a depth beyond the reach of most plant roots. The taller forests are forced to give way to wallum heathlands; stunted in appearance yet rich in species.

Because this high diversity of vegetation types is still intact, visitors are able to see one of the most remarkable and diverse sets of plants and animals in the world.

Ghost crabs inhabit dunes from the beach to about 200 metres inland. As their name suggests, these small, light-coloured crabs can be elusive and well camouflaged. Sharp-eyed campers and walkers may spot them as they scurry out of view. The crabs are most active at night, preferring to hunt and scavenge under the cover of darkness. During the day, like so many other dune animals, they retreat to their burrows. Here, they find shelter from the exposed, dry conditions and daytime predators.

Giant earthworms are one of the island's most impressive native invertebrates, some reaching 80 centimetres in length. They tunnel into the dunes by successively contracting each muscle along their segmented bodies using wave-like motions. Being hermaphroditic (both male and female), the worms have done away with the need to find a mate. After cross-fertilising themselves, they lay cocoons from which the juvenile giant earthworms later emerge. In general, very little is known about the island's worm fauna, which, although an integral part of the sand ecosystem, appears to be limited in diversity.

The giant burrowing cockroach is another land-dweller of sizable proportions. Reaching a length of up to 80 millimetres and weighing in at a whopping 35 grams, it is just one of hundreds of harmless native cockroaches known to inhabit the Australian bush. On the mainland, the few

*Foxtails - a common understorey plant*

# Introduction

species considered household pests have all been introduced from overseas (including the inappropriately named 'Australian' cockroach).

Cicadas are commonly heard on the island. At times, the sound produced by these large-winged bugs can be almost deafening. The males, which are responsible for the constant ringing, use specialised drumming and amplifying apparatus.

The larvae of scribbly (or inscripta) moths are responsible for the intriguing zig-zag patterns on the trunks of scribbly gums - a type of smooth-barked eucalypt. Before the tree sheds its old bark, the larvae tunnel into the junction between the old and new season's bark, leaving trails of intricate embossed markings that are worth a closer look.

Equally fascinating, but often mistaken for small flies, are the island's native bees. Bees play an integral role in maintaining healthy ecosystems since, without regular visits from these tiny nectar gatherers, many native plant species would not be able to cross-pollinate and set seed. The native bees live in colonies thousands strong, usually in tree hollows or fallen logs. During the day, they may gather

*Top to bottom: midyim berry, native iris, Wide Bay boronia*

*Below: centre: scribbly gum markings, right: leaf litter offers vital habitat*

*300 ant species inhabit the island*

50

# Flora & fauna

nectar from a variety of plants including grass trees, wedding bushes and large yellow guinea flowers.

At least twenty percent of Australia's 1,500 native ant species are thought to reside on Fraser Island. Given the sheer size of the task, it's not surprising that taxonomists face a big challenge when it comes to identifying and classifying species. Ants play important roles in the island's ecology; their impressive species diversity accommodated by the abundance of sand, complex soil processes and different vegetation types. Many species of ants are known to disperse plant seeds and trigger germination. Scientists are beginning to appreciate their complex associations with fire and suggest fire may be necessary to the long-term survival of the island's heathland colonies.

While dingoes are perhaps the most obvious, at least 47 other species of land mammal inhabit Fraser Island. Many are small, nocturnal creatures, like native rodents and insectivorous bats. Although feral cats have been recorded on the island, the absence of foxes may explain why so many small mammals have managed to survive. Other native mammals include echidnas, bats, wallabies, potoroos, dasyurids (small carnivorous marsupials),

*Top to bottom: Guinea flower, blue flax lily berries, goanna*

# Introduction

possums and gliders.

The long-nosed bandicoot, a small solitary marsupial with a long, pointy snout, uses its front feet to scratch out small conical holes in the ground as it forages for insects and other foods at night. Its shrill, grunt-like squeak is sometimes heard in between diggings as it actively sniffs and moves about.

Gliders inhabit the trees above. These appealing marsupials share the same superfamily as possums. Adults reach just over 20 centimetres in length. Special membranes extending from their fifth fingers to their first toes help them to glide for distances of up to 50 metres.

Sugar gliders are sought after by animal spotlighters but are only occasionally seen making the incredible glides for which they are famous. The best way to spot one is to listen for faint crackling sounds and dropping leaves as it feeds on sap, nectar or insects in the foliage above. (If there is a lot of noise, the animal responsible is probably a fruit bat).

Tree hollows make ideal nests in which groups of gliders are known to huddle together in cold weather. Although mostly social animals, with groups consisting of up to seven adults with their offspring, gliders generally spread out within their range while foraging. Ironically, it is thought there are more sugar gliders kept as pets in the United States than occur naturally in the Australian bush.

Feathertail gliders are more elusive. Adults grow to a mere 80 millimetres in length with a feather-like tail that makes the species unique amongst all Australian mammals. These tails add about the same length again and are crucial to gliding, steering, braking and anchoring. In spite of their small size, feathertail gliders can cover distances up to 20 metres in one glide hence the name *Acrobates pyg-*

*Grey-headed flying foxes*

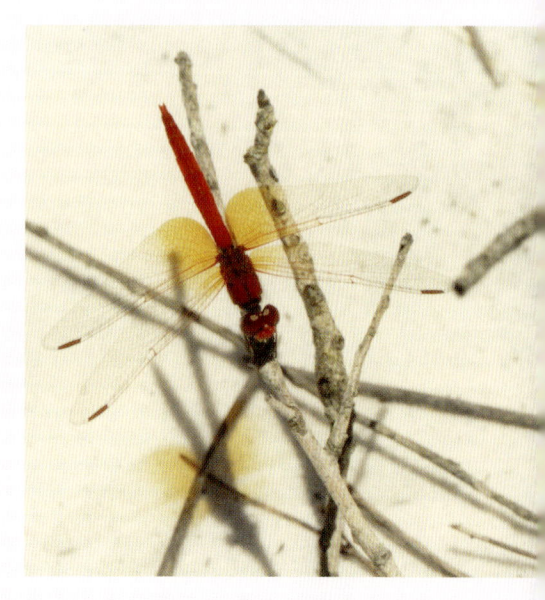

*Dragonfly at the water's edge*

# Flora & fauna

*maeus*. Like sugartails, they are sociable animals, often constructing their nests in hollowed-out tree limbs. On rare occasions, they have been seen congregating in the dozens in single trees, leaping and scampering from limb to limb in a mysterious frenzy of excitement.

Bats are among Fraser Island's most prominent nocturnal animals and play a vital role in plant pollination and insect control. This often-forgotten group makes up about a quarter of all Australian mammals. At least 19 different species visit or reside on the island, including fruit (flying foxes), blossom, sheathtail, mastiff and simple nosed bats.

To access the island's abundance of food, many bats commute nightly from daytime roosts in Hervey Bay. Others may travel from as far as Gympie. When the eucalypts are in flower, the noisy squabbles of fruit bats, like the grey-headed flying fox, are often heard in the trees above. Small insectivorous bats demonstrate superb in-flight agility using ultrasonic echolocation. In a single night, one of these bats can consume half its body weight in insects.

## Life in the lakes & swamps

Described by scientists as unusual and distinctive, little is known about the invertebrates of Fraser Island's lakes. The few species able to survive in the acid environment are either highly specialised, even unique to the lakes, or tolerant and comparatively widespread. The former includes an unusual non-biting midge that is providing insights into the diverse family to which it belongs.

Dragonflies are often seen around the edges of lakes and swamps and rely on the presence of water in order to breed. A much closer look can reveal a range of other aquatic insects as well as water beetles and bugs. Aquatic snails and slugs, on the other hand, are absent from the lakes and scientists have only recorded one species of leech, often found attached to the gills of freshwater fish.

Frogs abound on Fraser Island, making frog spotting a rewarding nightime pursuit. At least 17 native species inhabit the island's lakes and swamps. Sometimes mistaken for crickets or cicadas, each emits its own distinctive call, such as 'wark', 'aaaare', 'reeeek' or 'bonk'. (Only

*Boomerang Lakes*

# Introduction

*Sword sedge in seed*

*Paperbark*

the local green treefrogs sport a more traditional sounding croak).

Male frogs are especially vocal when it rains and during their breeding seasons. Many sing collectively in an effort to attract mates. Try the 'triangulation method' of spotting them at night. This involves two or more of you quietly gathering around the calling frog. Each person shines their torch in the direction that they think the particular call is coming from. The calling frog should be situated at the point where the torch beams intersect.

Tiny sedgefrogs can often be spotted clinging to the sedges around the edges of swamps, lakes and fens. Just above, emerald-spotted tree frogs are perfectly camouflaged against the branches of overhanging paperbarks. Other species include the aptly-named wallum rocket-frog, naked treefrog, beeping froglet and scarlet-sided pobblebonk.

The low pH of the island's wetlands cre-

*Kreft's river turtle, Lake Allom*

# Flora & fauna

ates favourable conditions for several species of 'acid frog'. Three of these have a limited range - the Cooloola sedgefrog, wallum sedgefrog and wallum froglet. Often regarded as environmental indicators, these acid frogs are highly sensitive to non-acid conditions and any upsets to water levels and quality. The island's acid habitats are incredibly important, offering some of the 'last bastions' of healthy acid frog habitat. On the mainland, vast tracts of coastal wallum have been severely degraded or drained and cleared for development along the southern Queensland and northern New South Wales' coasts.

Snakes, like tree snakes, carpet pythons and freshwater or keelback snakes, prey on frogs and other aquatic wildlife. While these particular species are non-threatening to humans, venomous snakes are occasionally seen in the vicinity of creeks, lakes, other water bodies and elsewhere.

It is not clear how the presence of cane toads may have impacted on local frog, snake and lizard populations. Originating from Central and South America, this invasive frog species is currently thought to be wreaking havoc on many of Australia's subtropical and tropical wetland ecosystems. Cane toads were deliberately introduced to the cane fields of far north Queensland in 1935 in order to control cane beetles. However the beetles survived and the toads flourished. How the toads made their way onto Fraser Island is a point of speculation but their hardiness would have made it easy for them to hitch a ride on floating debris or materials being transported across the strait.

While some island species, like keelback snakes, can reportedly avoid being poisoned by the toad's toxin, cane toads are suspected of displacing or poisoning many native species on the mainland. The red-bellied black snake, for example, has plummeted in numbers in areas in

*Native lily, Ocean Lake*

which it once thrived.

Mature cane toads are very large, reaching 20 centimetres in size, with leathery, warty skin and enlarged poison glands. They lay long strings of eggs, often in stagnant pools of water and, unlike native frogs, the tadpoles feed and swim in schools. Juvenile toads, however, can be difficult to distinguish from some native frogs so their destruction is not advised. Toad-busting activities are occasionally held on the island and help to educate people about responsible, humane ways of controlling the pest. To date, effective eradication measures have not been identified or carried out.

Apart from snakes, other reptiles that inhabit the island's wetlands include goannas, freshwater turtles and water dragons. Gould's goannas and lace monitors are often spotted and if disturbed, quickly seek refuge in a burrow or up the trunk of the nearest tree. Both hunters and scavengers, these large lizards patrol picnic grounds and raid turtle and birds' nests for easy meals. Visitors to less frequented lake beaches occasionally spot diggings and broken egg shells.

These shells usually belong to freshwater turtles. Three species inhabit Fraser Island - the broad-shelled, long-necked and Kreft's river turtle. Although the acidic lakes they inhabit are relatively devoid of

# Introduction

plant and animal life, these creatures survive by feeding on insects that fall on to the water's surface. In times of hardship, they are capable of moving overland in an effort to relocate to another lake.

The best known of these, the Kreft's river turtle, is often seen at lakes Allom, Bowarrady and Basin. Distinguished by a prominent pale yellow streak along the side of its head, it is a type of short-necked turtle belonging to the Chelidae family. Family members, which include the Brisbane short-necked turtle, typically have webbed, clawed feet and can tuck their heads in sideways. Scientists suspect the island's Kreft's river turtles might in fact represent a new species, quite different to their mainland counterparts.

Freshwater fish inhabit most of Fraser Island's lakes, with as many as 13 species occurring in Lake Wabby. Scientists speculate this may be due to Lake Wabby's depth, link with the water table and association with the sea. In total, at least 20 native species and one introduced species are recorded for the whole of the island. While most of the lakes support only one or two species, these fish play an important role in freshwater ecology.

Long finned eels, some reaching well over a metre in length, are often seen in many of the swamps and creeks, including Eli Creek and Wanggoolba Creek. As part of their fascinating life cycle, juveniles make the long journey from their birthplace in the Coral Sea down the east coast of Australia to seek out water bodies in which to grow and mature. In spite of their ability to move over land, many never make the return journey and instead become trapped in lakes and dams.

The southern or soft-spined sunfish, a type of rainbowfish, is at home in soft water bodies with sandy bottoms and inhabits many of Fraser Island's lakes. Males, in particular, may develop deep red or bright blue colouration, perhaps as a response to environmental conditions, predatory fish or interactions with rival males. Although endangered on the mainland, where coastal development has destroyed most of its wallum habitat, southern sunfish are believed to be locally abundant on Fraser Island.

Another endangered fish, the honey blue-eye, inhabits Lake Wabby, the Noosa River and a limited number of other suitable water bodies between Brisbane and Bundaberg. Honey blue-eyes feed on various aquatic invertebrates and microscopic organisms. They spawn in submerged vegetation, where the fertilised eggs, which attach to plants, later give rise to surface-feeding juveniles.

Firetail gudgeons can be seen in many of the lakes and creeks, including Lake Birrabeen. In the lead up to breeding, the male of this comparatively common species develops a large bulbous growth on his head. Just a few centimetres in length, his body is usually bronze to grey in colour, with pale red dorsal margins and anal fins.

Other freshwater fish include purple spotted gudgeons, carp gudgeons, crimson-spotted rainbow fish, olive perchlets and flyspeckled hardyheads. Little is known about the endangered Oxleyan pygmy perch whose abundance may be locally limited.

Although their habitat is protected on Fraser Island, these often-isolated fish populations may be vulnerable to chemical pollutants and subtle changes in water quality. Exotic mosquito fish are present in many of the water bodies and may compete for food, prey on eggs and behave aggressively towards the native species. Visitors must play their part in helping to protect the island's lakes, streams and wetlands by ensuring detergents, sunscreens, insect repellents and

# Flora & fauna

# Introduction

other chemicals are prevented from entering the water. Fishing is not permitted in any of the island's lakes or creeks.

A huge variety of plants flourish in the swamps and around the edges of lakes. They form a range of different plant communities, unique associations and microhabitats; all vital for sustaining the diversity of aquatic, terrestrial and avian animal life. Many species, like the small carnivorous sundews, have special adaptations to help them overcome the challenge of obtaining nutrients needed for plant growth. The vegetation changes dramatically across the island - creeks and lakes spring beside densely vegetated dunes and towering forests suddenly give way to low, stunted sedgelands.

In times of drought, some of the lakes transform into desolate beds of parched, cracked soil but even here, it is possible to see a diverse mix of survivors and colonisers. This is especially noticeable around the edges of lakes that receive very few visitors. The seedlings of many different plants, including banksias, paperbarks and sheoaks, emerge in every available space, while sedges and rushes flourish in the mud.

All of the island's lakes are characterised by distinct zones of vegetation running parallel to shore. This natural phenomenon often begins with rushes in the shallows, a mix of pioneer species on the beach and clumps of sedges and rushes on the white sand behind. These pioneer zones tend to be followed by bands of flowering heath, paperbarks, shrubs then denser eucalypt or banksia woodlands. The different zones reflect changing conditions as the distance from the water increases. Closest to the edge, fluctuations in water level play an important role. Moving away from the shore, a range of factors such as soil moisture, wind exposure, soil nutrients and the presence of particular plants and animals may influence which species grow.

Many of the island's larger swamps are a breathtaking sight. In the late afternoon, the sun often basks them in a soft golden light. Across the wild expanse of

# Flora & fauna

Wathumba Swamp, for example, a profusion of flowering shrubs and orchids emerge from the seemingly endless sea of sedges. It's easy to imagine that healthy wetlands like this are literally teeming with life.

Swamp wallabies occur on the island but are only very rarely seen by visitors. The reason for this, at least in part, is that they prefer swampy areas with dense undergrowth that are less frequented by dingoes (and people). These attractive, coarse-haired wallabies are recognisable by their brown and rufous orange-coloured coats and darker extremities. The species is so distinct from all other wallabies that it inspired its very own genus, *Wallabia*.

Swamp wallaby joey

# Introduction

*Soldier crab*

## Sand flat & shallow dwellers

The Great Sandy Strait is a wetland of international importance (RAMSAR). Intertidal mudflats make up a third of it, with the remainder a mix of mangroves, seagrass, saltmarsh, sandspits and small islands. Many fish depend on these areas, particularly the mangroves and creeks, for parts of their life cycles. Young mullet, for example, move into freshwater streams for part of their early development, while jungle perch and Australian bass seek out brackish water in order to breed.

In the intertidal zone of the island's surf beaches, eugari or 'pipis' (a triangular-shaped mollusc) are commonly found just beneath the sand's surface. They are collected for bait and human consumption (limits and seasonal restrictions apply) and were once a staple food of the Butchulla people. Mounds of their bleached shells, some thousands of years old, reveal traditional eating areas. Some of today's Butchulla descendants consider the impact caused by unrestricted traffic along the eastern beach to be disrespectful towards this precious resource.

Walking is a great way to enjoy the beaches. As the tide falls along the island's west coast, walkers have a unique chance to explore the rich sandflats. When the time is right, soldier crabs

*Evidence of crabs sieving for food*

# Flora & fauna

emerge in their thousands, forming impressive, brightly-coloured 'armies'. To the delight of onlookers, especially children, they make characteristic 'tinkling' sounds as they move around en masse. As with many other animals, this grouping behaviour is thought to confuse predators.

The crabs feed using their claws to scoop up mouthfuls of sand, which they sieve for edible matter. Each crab leaves a trail of small pellets of discarded sand. Adult soldier crabs are small in size, with globular, sky-blue bodies. Their ability to walk forwards sets them apart from other species of crabs. When danger approaches, they burrow into the sand using a circular corkscrew motion.

Another pellet producing crab is the sand bubbler, a not-too-distant relative of the soldier crab. The crabs eject tiny balls of sand as part of their feeding and 'housekeeping' activities which radiate out from the central burrow in a distinct star-like pattern. Sand bubblers differ from all other crabs in their ability to absorb oxygen from the air through special patches on their legs called tympana.

Fiddler crabs also emerge at low tide from the mangrove mud near the mouth of freshwater creeks. Male fiddlers are easy to spot; each has an oversized orange coloured claw used to fight and intimidate other males. He also waves it around to attract the attention of females. The males can only eat using their smaller claws, and so must devote more time to feeding than the comparatively dexterous females. As the tide comes in, each crab returns to its burrow, sealing it tight with a mud plug and bubble of air.

Mudskippers are another common burrow dweller, also active at low tide. They are an unusual type of fish in that they can leave the water in pursuit of prey. Thanks to specially modified pectoral fins, mudskippers can propel themselves along with surprising speed and agility. They can breathe both in and out of the water as well using their gills and other parts of their bodies. When danger approaches, they quickly retreat into their burrows.

The mangrove forests come alive with various animals and invertebrates at both high and low tide. Often unrelated,

*Mangrove habitat, Wathumba Creek*

# Introduction

plants described as 'mangroves' share the unique ability to cope with fluctuations between salt and freshwater, exposure to hot, dry conditions and highly anaerobic (oxygen depleted) soils. Because of their importance to the ecosystem, all mangroves are protected by law. One of the most common species, the grey mangrove, is easily recognised by its short pen-like roots that emerge like snorkels from the surrounding sand.

An incoming tide brings marine excavators into the shallows, especially along the western shore. Stingrays and shovelnose rays are responsible for most of the large holes that riddle the beach at low tide. Shovelnose rays (also known as fiddler rays) are considered harmless to people. As their name suggests, their snouts are sharply triangular in shape; all the better to sift sand with. Their mottled brown colour makes them difficult to spot, although some nightwalkers get a good look when shining their torches off the Kingfisher Bay jetty. The tails of stingrays, on the other hand, are lined with venomous barbs, which are used to ward off predators and defend themselves against attack. The golden trevally, a prized sport and eating fish that grows up to a metre long, is also known to excavate small, deep depressions close to shore.

## Creatures of the deep

Fraser Island's offshore waters are rich in marine life due, in part, to their geographic position and overlap with the Great Barrier Reef's most southern limit. These waters provide a vital habitat for a number of vulnerable and endangered species, including six species of marine turtles, the humpback whale, dugong and Indo Pacific humpback dolphin. One of the most interesting groups to visitors - the marine mammals - is well represented with at least six species of dolphin, two species of pilot whales, dugongs and five other species of whale recorded for the area. Although rarely seen, the Australian sea-lion and New Zealand fur-seal also make the list.

Platypus Bay is famous for its humpback whales, which use its warm, sheltered waters as a resting place between July and November each year. Covering some 5000 kilometres, these large mammals migrate from Antarctica northwards to the Whitsundays in order to breed and suckle their calves. Although forced to the brink of extinction by whalers last century, numbers are thought to be on the rise. Unlike many other whales, humpbacks sometimes display vigorous behaviour at the water's surface. Thousands of tourists take to whale-watch vessels each year to witness this spectacle off the island's northwest coast. On the other side of the island, binoculars may be needed to spot the blows and humps of whales as they pass by further out to sea.

Sea turtles are sometimes spotted from passing boats and off vantage points like Indian Head. In certain places, like Rooney Point, these prehistoric-looking reptiles cruise up and down the sheltered channels close to shore, making them easy to spot from the beach. Hervey Bay offers an important courting and mating ground for loggerhead turtles. On the mainland, a few hundred females come in to nest at Mon Repos each year. Green and leatherback turtles also nest in the area and among the other recorded species are the hawksbill, flatback and Pacific Ridley. All are endangered, vulnerable or potentially vulnerable.

Hervey Bay's extensive seagrass beds provide a vital food source to the area's surviving dugongs (or seacows). An adult dugong devours 30kg of seagrass each day. One or more animals are occasionally spotted along the island's south west coast, appearing as large, slow-moving

# Flora & fauna

shapes that only surface once every five to ten minutes. Dead or stranded dugongs and any unusual sightings should be reported to authorities.

Already extinct in some island waters of the Pacific, dugongs between here and Cooktown are thought to have halved in number since 1980. Only a few thousand may be left on the planet. Among other things, the seagrass beds upon which dugongs depend are vulnerable to damage from boats, agricultural runoff and sedimentation caused by erosion and flooding. Two cyclones in 1992 wreaked havoc on them, causing local dugong numbers to plummet. Furthermore, although long-lived (up to 70 years), dugongs are slow to mature. At best, a fertile female might give birth to one calf every 3-7 years.

The variety of offshore habitats supports a high diversity of open water, rock and reef fish. Thanks to Fraser Island's contribution, south east Queensland is able to boast nearly one third of Australia's total number of marine and estuarine fish (well over 1000 species). Along the west coast, extensive sandflats and mangroves provide important breeding grounds for many of the saltwater species.

Tuna and mackerel may be seen schooling near the water's surface off the northwest coast, while bream, whiting and flathead are commonly caught off the beaches. Tailor is probably the most well-known species of fish caught off the island. By September each year, Fraser Island and Cooloola's surf beaches support annual spawning runs of tailor. This attractive silver coloured fish, said to reach up to 14 kilograms in weight, was once a seasonally significant food resource for the island's Butchulla people and visiting tribes.

The same deep gullies that bring tailor close to shore provide the same opportunity for sharks. Tiger and mako sharks are often sighted. Grey nurse sharks reach their northern limit and are a protected

*Humpback whales*

# Introduction

*Avoid treading on washed up bluebottles*

species. Other species include wobbegongs, tiger sharks, hammerheads and dusky whalers.

Sea snakes are well represented in the ocean waters off Fraser Island. Strandings of at least 11 different species have been recorded, primarily along the eastern beaches. Many are highly venomous so keep well away. Little is known about these beautiful snakes, their numbers, movements or the roles they play in marine ecosystems. The yellow-bellied sea snake feeds on the surface in open waters whereas olive sea snakes are found 5-20 metres below.

## Animals to be wary of...

Like most other natural areas in Australia, Fraser Island is home to a range of animals that can potentially threaten the safety or comfort of humans. Only a few examples are provided here. Visitors with animal phobias and medical conditions, such as insect allergies, may need to make inquiries before making the decision to visit. Incidences involving animals such as dingoes should be reported to QPWS staff and medical advice should be sought promptly for injuries or symptoms that raise any concerns. Fraser Island is remote and, on some parts of the island and at certain times, medical help can be many hours away. It is a good idea to carry a well-stocked first aid kit and be familiar with first aid techniques.

When you encounter wildlife, common sense is the key. It is always best to err on the side of caution. Keep well away from potentially threatening animals, taking care not to aggravate, handle, threaten or disturb them in any way. The island and its offshore waters support a number of venomous species, including snakes, spiders and marine stingers, but serious accidents involving such creatures are rare, compared with the number of dingo-related instances for example. (See p 44-47 for information on dingoes - one of the main species to be wary of).

Some spiders will bite readily if provoked and, in the case of the Fraser Island Funnel-web, the venom can be highly toxic

# Flora & fauna

and potentially life threatening. Other spiders, like huntsmans, redbacks and daddy-long-legs, which are common in many Australian homes and gardens, sometimes cause psychological distress to overseas visitors. While redback bites are highly variable in their effect, they are potentially life-threatening, requiring early medical attention. Huntsmans are often timid, with bites generally producing mild, localised pain. Daddy-long-legs, on the other hand, are completely harmless to humans but ironically are able to kill huntsmans and redbacks. Dozens of different species of spider occur on the island, many of which are new to science. Interested or concerned visitors should seek advice from local ranger staff.

Although not specific to the island, there are a number of excellent field guides with pictures and detailed information on Australian wildlife and dangerous animals. Most book stores stock these.

Biting insects, like sandflies (midges) and mosquitoes, can be notoriously bad on Fraser Island, causing itching and discomfort for many unprepared visitors. They can trigger more severe reactions in people who have allergic predispositions.

Sandflies, for example, can inflict hundreds of bites to exposed skin in a matter of minutes. These tiny insects are most abundant around the creeks and mangroves along the island's west coast, especially during the summer months. It is advisable to cover up or apply repellent before arriving in these areas to avoid being bitten.

Mosquitoes are also common to the wetter parts of the island, including the interior rainforests. A little way back from the beaches, march flies can at times seem relentless in their harassment of people. After landing on skin or fabric (many locals claim they are attracted to blue-coloured clothing), the large biting fly will pause momentarily before inserting its proboscis. The bite usually leads to a mild swelling, similar to a mosquito bite.

Although effective against these insects, insect repellents are best avoided if you are planning to swim in the island's lakes. The chemicals they contain pollute the water, impacting on native insects, fish, frogs and other sensitive species. Many people opt to kill persistent, slow-moving march flies by hand and clothing will generally deter other biters.

If swimming or wading in salt water, keep well clear of any suspected blooms of *Lyngbya*, commonly referred to as fireweed on Fraser Island. *Lyngbya* is a kind of marine bacterium that can sting and cause irritation upon contact. The severity of a person's reaction generally depends on their level of exposure. Large floating mats or 'blooms', which are usually red-brown in colour, can occur along the beaches close to shore, as well as in the Champagne Pools. People are advised to avoid coming into contact with floating, growing or washed up *Lyngbya*. Symptoms include stinging, burning, itching, swelling and, in severe cases, blistering and respiratory problems. The latter may result from breathing in dried algal matter. First aid measures recommended by QPWS include washing any affected skin with soap and water, flushing eyes out with fresh water (if affected) and applying cool compresses if the irritation persists. Medical attention is advised if the person's eyes are affected, the extent and severity of their irritation is causing concern or if they are experiencing respiratory discomfort. Since avoidance is the key, it can help to ask island locals about fireweed outbreaks when you arrive.

Poisonous animals sometimes wash up onto the beach, particularly the surf beaches. Thousands of bluebottles litter the beach at certain times of the year

# Introduction

(also known as Portuguese Man-O-War). This small type of ocean jellyfish has venomous tentacles that it uses to ensnare fish. In the water, these are suspended by a clear blue elongated float, several centimetres in length. North-easterly winds blow the jellyfish onshore, where people without footwear are at risk of treading on them. Contact with the tentacles' stinging cells can cause severe local skin pain as well as other symptoms. The application of cold packs is recommended and medical assistance may be necessary.

Sea snakes are occasionally washed up and should be treated with extreme caution. Although a few species are thought to behave placidly in their natural surrounds, most sea snakes are venomous and many are described as probably being capable of delivering a fatal bite (see p 64 for more information on sea snakes).

In calm shallow water, it pays to be wary of estuarine stingrays, a type of large, flat bottom-dwelling fish that feeds in intertidal areas. If threatened or trodden on, a stingray can inflict an excruciatingly painful wound with one swipe of its tail. Likewise, the much smaller stingaree, which also has stinging barbs along its tail, is best avoided.

On land, particularly in picnic areas and campgrounds, visitors need to watch for scavenging animals. The largest member of the kingfisher family, a bird known as the kookaburra, is notorious for snatching food from people's hands (even mouths) and can inflict injuries when attempting to do so. Goannas may also climb onto picnic tables and go to some lengths to obtain food - another reason why visitors must take great care to keep food and rubbish securely stowed.

Magpies may swoop aggressively at people during their breeding season. Most Australians are familiar with the antics of this charismatic bird, which only rarely result in injury. The swooping is intended to ward off intruders so the best thing to do is leave the area.

Bushwalkers, in particular, should continually keep an eye out for snakes that may be crossing the track or sunning themselves across the path. In virtually all of Australia's natural areas, it pays to watch where you tread, whether walking through bush, grassland, dunes or other habitat types. In addition to supporting several species of pythons and tree snakes, Fraser Island has at least 13 species of front-fanged snakes including the highly venomous coastal taipan, southern death adder, red-bellied black snake and eastern brown snake. Most are considered dangerous or potentially dangerous but in most cases, may only bite if cornered, trodden on or otherwise provoked. Moving snakes that are given a wide birth will usually go quietly on their way. QPWS suggests walkers watch for snakes, wear sturdy boots, use a torch at night and leave snakes alone.

Pythons, tree snakes and keel-back snakes, which are not considered dangerous, are often sighted, but can sometimes be difficult to distinguish from the venomous species. Incidences of bites on the island are very rare. The standard first aid procedure, in the event of a bite or suspected bite, is to immobilise and apply a pressure bandage to the limb (leaving any residue at the bite site on the skin for later identification) and seek urgent medical attention (also applicable to sea snake bites).

Cane toads have large glands on the sides of their heads that can exude a toxic substance, especially if pressure is applied to them. Visitors should therefore avoid handling toads or prodding them with implements.

In the northern part of the island, especially around Orchid Beach and Waddy

# Flora & fauna

Point, visitors are advised to keep well clear of the small herd of wild horses. Although seemingly docile and tame, brumbies can sometimes kick or bite.

Another type of mammal not often thought of as dangerous is the humpback whale. In rare instances, whales have become entangled in anchor chains or otherwise felt threatened and displayed aggressive behaviour towards boats. All boats are required by law not to approach within a certain distance of any whales and to turn off any motors at this distance. Whales may, of course, decide to venture nearer and in such instances, are almost always curious, rather than aggressive. The stationary vessel must remain adrift until the animals have moved off. Stranded whales can also present hazards to well-meaning onlookers. Strandings should be reported to QPWS rangers and rescue attempts by inexperienced persons are not recommended.

High numbers of sharks are known to patrol the waters off Fraser Island. Their dark outlines can sometimes be seen as they swim very close to shore in pursuit of fish or other prey. This, combined with the fact that there are numerous rips and no life guards, make the island's surf beaches, in particular, a potentially dangerous place to swim. Swimming is therefore not recommended.

*Brumbies, Waddy Point*

67

# Northern Fraser Island

Some of the island's most remote destinations are found in the north.

Here, you can experience wild, windswept scenery - from stunning beaches and towering dunes to rocky outcrops and the spectacular Champagne Pools. For more serene surroundings, check out Ocean Lake or cross the island's scenic heathlands to magnificent Wathumba Creek.

Exploring the island's north is also a great way to escape the crowds. Generally speaking, the further north you go, the less people you are likely to encounter.

For a safe journey, make sure you have plenty of time, are fully prepared and pay close attention to the tides. Many drivers choose to travel in convoy.

*Sandy Cape*

*Ngkala Rocks*

*Wathumba Creek*

68

SANDY CAPE
SANDY CAPE LIGHTHOUSE
CORAL
SEA
FRASER
ROONEY POINT
PLATYPUS
BAY
OCEAN LAKE
MARLOO
BAY
WADDY POINT
ORCHID BEACH
WATHUMBA CREEK
CHAMPAGNE POOLS
INDIAN HEAD
GREAT SANDY
AWINYA CREEK
NATIONAL PARK
WARRADY CREEK
LAKE GNARANN
WHITE LAKE
ALIE CREEK

# Northern Fraser Island

**Map reference**: A9 (Hema), A7 (Sunmap)

**Location**: 26 km from Ocean Lake, 31 km from Orchid Beach, 39 km from Indian Head

**What's there?** Extensive beaches, excellent fishing, sense of isolation

**Nearest to**: Ocean Lake, Rooney Point, Sandy Cape Lighthouse

## Why go?

Sandy Cape has its own unique sense of space and isolation, offering a remote, wilderness experience ideal for those serious about fishing or simply wanting to escape the crowds.

Breaksea Spit, which extends more than 30 km north, attracts fishing enthusiasts. Tailor, dart, bream, trevally and whiting are commonly caught off the beach.

Submerged wrecks such as the Seabelle (wrecked 1857) and Chang Chow (wrecked 1884) occupy the waters off Sandy Cape. The wrecks are testimony to the numerous ships sacrificed in the area to hidden sandbars, changing tides and weather extremes. It was in this region that Captain James Fraser and his wife, Eliza, came ashore in a longboat, after losing their ship The Stirling Castle to the Great Barrier Reef further north.

## Getting there

Time and planning are needed to get to Sandy Cape. The area is semi-remote, and can only be accessed from the island's eastern beach so drive as near to low tide as possible and allow plenty of time. When driving on and off the beach, keep to obvious tracks. Some bypasses such as the one at South Nkgala Rocks can be soft and steep, presenting difficulties for some drivers, particularly those lacking sand driving experience (see Driving on Fraser Island p 22-35). Extreme care should also be taken when tackling rocky sections; a high clearance vehicle is advisable. Most hire cars are not permitted beyond Waddy Point (confirm this with your hire company).

## Facilities

Sandy Cape has no facilities. The nearest facilities, including fuel, general supplies and a payphone, are located at Orchid Beach. Camping is free-range. Try to use existing sites, taking care to avoid turtle nesting areas and damage to dune veg-

# Sandy Cape

*Looking out across Breaksea Spit - a favourite spot with fishing enthusiasts - it's easy to see why these were treacherous waters for ships.*

etation. Take all rubbish away with you.

## What to look for

Some fascinating plants and animals inhabit this region. Leatherback and loggerhead turtles are sometimes seen close to shore. Give these creatures a wide berth, especially during their breeding season (November-March). People and lights can upset their behaviour, even their desire to lay eggs. (On the mainland, Mon Repos near Bundaberg offers information and guided tours of turtle nesting sites).

Stunted, windswept vegetation adds to the area's unique wilderness quality. Perhaps most striking is the seemingly endless, growing stretch of foredunes with their undulating hummocks and silvery grasses. Here, native spinifex grass sends out long runners to stabilise the dunes and trap any new sand; a clue that this part of the island is in fact growing.

Overhead, birds of prey such as brahminy kites, whistling kites, ospreys and sea eagles are often seen. A lucky observer may catch the unusual sight of wild horses (brumbies), although few remain.

Dingo skull studies suggest the island's northern dingoes might be its purest, possibly due to their isolation and territorial behaviour.

## What we think

Sandy Cape's reputation for difficult driving and unpredictable conditions leads many visitors to cross it off their itineraries - a plus if you're looking for peace and quiet or want to throw in a line. We love the feeling of space and remoteness here. So long as you plan ahead, allow plenty of time and drive sensibly, Sandy Cape should present few driving difficulties and plenty of great holiday experiences. For added peace of mind, why not get friends to "tag-a-long" in a second vehicle?

# Northern Fraser Island

**Map reference**: A8 / B8 (Hema), A6 (Sunmap)

**Location**: 7.5 km from Sandy Cape, 33.5 km from Ocean Lake, 46.5 km from Indian Head

**What's there?** Lighthouse, sandblow, walks, ocean views

**Nearest to**: Ocean lake, Rooney Point, Sandy Cape

## Why go?

Standing 26 metres tall and some 128 metres above sea level, the Sandy Cape Lighthouse looms impressively from atop Flinders Sand Blow.

Still in operation today, the lighthouse was built by the Queensland Government in 1870 to reduce the number of ships lost in the area. Twelve hundred tonnes of steel and concrete were hauled up for its construction, each panel of the lighthouse weighing 3.5 tonnes each. Whale and rapeseed oil fueled the first beam of light. Following further wrecks, responsibility for the operation of the lighthouse was transferred to the Australian Government in 1915. Fifteen years later, the rotating mechanism of the lighthouse was electrified. In the event of a breakdown, alarm bells would sound, alerting lightkeepers.

In 1995, the light was automated. Today, the lighthouse and 640-acre reserve is maintained by the QPWS Sandy Cape ranger (a former lightkeeper) with the help of volunteers. On a clear night, the beam from its solar-powered globe is visible for 28 nautical miles.

Whilst it's not possible to enter the tower, a number of historical exhibits can be viewed through the windows around the base. The historical details of the structure's construction and operation, as well as ships wrecked in the area, make for an interesting read. Ships aren't the only casualty though. In the 1960s, the sands of Flinders Sand Blow temporarily parted to reveal the mysterious wreckage of a plane, before covering it up again.

## Getting there

The lighthouse is situated just around the corner from the tip of the Cape. This 7.5 km drive should only be attempted in favourable conditions at low tide, particularly if you plan to walk up to the lighthouse. (For information on how to reach Sandy Cape, see p 70-71).

Three short walks, the start of which is signposted, begin on the beach adjacent to the lighthouse. One of these leads up to the lighthouse (1.2 km one way) however note that self guided visitor hours are between 8 am and 4 pm. The second (1.3 km one way) leads to the graves of the first head lightkeeper and his daughter, while the third (600 m one way) takes you to the remains of a World War II bunker.

## Facilities

There are no facilities here or in the vicinity apart from some interpretive signs and picnic tables. You will need to park high up on the beach providing conditions, such as tides, permit.

## What to look for

The walk up to the lighthouse is quite

# Sandy Cape Lighthouse

*High atop a mountain of sand and dense vegetation, the Sandy Cape Lighthouse provides an interesting stopover for visitors to the wild, remote north.*

steep, requiring a good level of fitness. It affords great views, however, including a forested gully on the right, dotted with the odd cabbage tree palm. Bungwall ferns are abundant while native sedges and groundcovers can often be seen in flower. A picnic table is situated about half way up which looks out across the sand blow, over the top of callitris and sheoak forest, towards the distant ocean. Look for birds of prey like brahminy and whistling kites.

The walk to the bunker is relatively short and easy, passing through a forest of large paperbarks and blackbutts. (The mosquitoes can be fierce). This is a relic from the RAAF No. 25 Radar Station (1942- 45) set up to detect enemy aircraft and ships during the Second World War.

The walk to the graves has lots of ups and downs. It takes you along a fairly wide but overgrown track, affording glimpses of the lighthouse.

## What we think

Whether viewed from the beach below or the grounds above, the lighthouse and Flinders Sand Blow are an impressive sight. The walk to up to the lighthouse is only worth doing if you have plenty of time as well as energy to burn. It makes you wonder what life would have been like for the labourers and lightkeepers.

*Grave of first head lightkeeper*

73

# Northern Fraser Island

**Map reference**: C6 (Hema), B5 (Sunmap)

**Location**: 13.5 km from Sandy Cape Lighthouse, 21 km from Sandy Cape, 60 km from Indian Head

**What's there?** Remote sheltered beaches, hiking, boating, fishing, swimming, marine wildlife

**Nearest to**: Ocean Lake, Sandy Cape, Sandy Cape Lighthouse

## Why go?

Rooney Point is a remote, wilderness beach setting surrounded by idyllic blue-green ocean waters. The beach is great for walking, boating, fishing, observing marine wildlife or watching the sun go down. Self-sufficient visitors can escape the 'hum' of beach traffic, since the area can only be accessed on foot or by boat.

## Getting there

Until recently, visitors were permitted to drive along the beach south of Sandy Cape, however this section is now closed to beach traffic and the only access is by boat or on foot. Because there are now fewer visitors to what was already a remote location, careful planning and preparation are a must.

Experienced hikers can head south along the beach from the Sandy Cape Lighthouse; a distance of 13.5 km one way. This is a relatively easy walk at low tide, but expect to get your feet wet crossing Bool Creek. Preparations such as prior arrangements for transport to and from Sandy Cape are your responsibility. Make sure any driving corresponds with low tide and avoid the risk of leaving a parked car on any beach.

Before anchoring a boat off Rooney Point, check weather forecasts. North - westerly winds (more common during winter) can spoil the calm conditions by bringing short, steep waves up onto the beach. It's hard to imagine the transformation of these usually calm waters to rough, treacherous seas with strong winds and hidden sandbars - but this is known to happen. Just four years after the construction of the Sandy Cape Lighthouse in 1870, cyclonic weather conditions caused an American barque known as the Panama to become wrecked just north of Rooney Point. Even today, during extreme whether conditions, boats are occasionally damaged or sunk in the area.

If you are thinking about taking a swim, bear in mind there are no life guards or nearby help. Also, avoid swimming or wading in the water during blooms of *Lyngbya* - a kind of marine bacterium that can sting and cause irritation upon contact (see p 65).

## Facilities

There are no facilities at Rooney Point. Hikers and boaties must be completely self-sufficient, taking their own food, water, equipment, first aid and other requirements. The nearest facilities, including fuel, general supplies and a payphone, are located at Orchid Beach. Camping is free-range. Avoid fragile dune vegetation, turtle nesting areas and other potential habitat areas.

*Rooney Point's serene, aquamarine waters are hard to beat*

# Rooney Point

*Sheltered from south-easterly winds, Rooney Point's idyllic waters are home to a diversity of marine wildlife as well as the wreck of the Panama.*

## What to look for

Marine wildlife is abundant in the area and since the Point provides 270-degree ocean views, your chances of spotting something are good. If you can, take a pair of binoculars.

Whales can be sighted between July and October. Look for their distinctive 'blow' which can be visible for kilometres on a clear day.

Dolphins and dugongs may also pay a visit and marine turtles sometimes swim close to shore, appearing to check you out. In the event of a female turtle coming ashore to lay her eggs or if hatchlings emerge, keep well away.

Beach and small boat fishing opportunities are good, with dart, bream, whiting and trevally caught off the beaches and a variety of reef fish just offshore. Tuna and mackerel are sometimes seen schooling near the water's surface in season.

Dingoes may be encountered in the area and like elsewhere, caution is advised. Don't leave food or rubbish unattended.

## What we think

Quite a few places on Fraser Island embody our notion of 'paradise' and this has to be one of them. The calm, crystal-clear waters are stunning, and there is a great opportunity to spot the odd turtle or dolphin. A distinct lack of people, revving vehicles and biting insects are another bonus, making this place a tempting alternative to the similar environment of Moon Point further south.

But come prepared for the likelihood of total isolation (even in peak holiday seasons) and don't be rushed for time - the island's north is not for anyone in a hurry. Above all, leave everything as you find it, taking all rubbish with you.

# Northern Fraser Island

**Map reference**: F9 (Hema), D7 (Sunmap)

**Location**: 5 km from Orchid Beach, 9 km from Waddy Point, 13 km from Indian Head, 21 km from Wathumba Creek

**What's there?** Freshwater lake, abundant birdlife, aquatic plants, lookout, easy walks

**Nearest to**: Orchid Beach, Waddy Point, Champagne Pools

## Why go?

The most accessible freshwater lake in the island's north, Ocean Lake is a bird watcher's paradise and a tempting destination for anyone interested in kayaking, photography, swimming or picnicking.

Visitors can enjoy panoramic views by taking the short walk to the lookout or marvel at the handiwork of local aquatic spiders in the forests fringing the lake.

As a window lake, Ocean Lake provides a window into the main water table below. Water flows from the lake towards the ocean via Orange Creek which is stained orange by tannin. In spite of this, the lake's water level remains constant and its naturally high nutrient load helps to support the plentiful birdlife. In fact,

*Webs spun by an unusual aquatic spider*

Ocean Lake is thought to have the island's highest population of breeding and nesting waterbirds.

A few people come to swim in the clean, slightly yellow water and make use of the small beach.

## Getting there

Situated north of Waddy Point, the only access by vehicle is along the beach. Keep an eye out for the small signpost marking the turnoff. In dry conditions, this track may become soft, requiring some momentum to get through. The lake and adjacent carpark is approximately 1 km inland, just past the bush camping area.

## Facilities

Day visitors are well catered for. Composting pit toilets, picnic tables and a large car park are situated close to the lake. Visitors with restricted mobility can gain relatively easy access to the lake. An interpretive board provides a comprehensive overview of the lake's ecology. Barbecues and fires are not permitted and soaps, detergents and shampoos must not be used in the lake. Cigarette bins are provided but water suitable for drinking is not available.

Whilst camping is not permitted near the lake, plenty of campsites are located near the turnoff from the beach. These sites cannot be booked in advance and no facilities are provided.

## What to look for

Long armed prawns, leeches, rocketfrogs, eels, and freshwater turtles inhabit the lake, while the surrounding forests are home to goannas, poisonous funnel web spiders and a strange aquatic spider that doesn't like water. Significantly, Ocean Lake is the only lake on the only island known to support carp gudgeon. Not only this, but the fact that these small fresh-

# Ocean Lake

*Flourishing aquatic plants create a unique, freshwater oasis sure to please bird watchers and nature lovers visiting the island's north.*

water fishes coexist with firetail gudgeons makes the lake unique in the world.

Birds are by far the most conspicuous creatures, with willie wagtails, butcherbirds, honeyeaters, and bar-shouldered doves commonly sighted. The abundant aquatic birdlife includes pied cormorants, black swans, musk ducks and pelicans.

Walkers have two choices, both of which finish up at the lookout. As well as the short lookout walk, the Cypress Circuit offers an easy 1km walk that begins with countless unusual spider webs hanging overhead. Laden with fallen leaves, these abandoned webs belong to an unusual type of water spider depicted in the interpretive signage.

Water lilies flourish in the lake while on land, sheoaks, white cypress, banksias, acacias and paperbarks grow with an understorey of hyacinth orchids, sedges and geebungs.

From the lookout, the odd cabbage palm and pandanas palm can be spotted in the forest below, with a panorama that takes in the nearby ocean. Furry-looking lichens grow on the cypress trees and this is the perfect vantage point from which to hear and see birds.

## What we think

Keen bird watchers could easily spend a day or two at Ocean Lake. Even the car park puts on a good show! Why not take a chair and find a quiet spot from which to observe the many water birds? Or if you have energy to burn, consider exploring the lake from a canoe or kayak. Paddling around the lilies is great fun.

Whatever your reason for visiting, Ocean Lake never fails to delight. The short walk to the lookout is an absolute must and well worth a detour, so long as you don't mind the spider webs overhead.

# Northern Fraser Island

**Map reference**: G10 (Hema), D7 (Sunmap)

**Location**: 2 km from Champagne Pools, 5 km from Waddy Point, 8 km from Indian Head

**What's there?** Small seaside village with shop & holiday houses, fishing

**Nearest to**: Champagne Pools, Ocean Lake, Waddy Point

## Why go?

The most northern settlement in the island's remote north, Orchid Beach and its well-equipped general store is a popular stop for fishing enthusiasts, Waddy Point campers and people heading north to explore Ocean Lake or Sandy Cape.

Originally developed as a tourist resort in 1963 (which was later condemned) then controversially subdivided in the 1980s,

*Hyacinth orchid*

Orchid Beach now the site of some of the most expensive real estate on the island. Dozens of large houses are nestled in the dunes, with magnificent views across Marloo Bay. Most are holiday rentals, their bitumen roads and manicured lawns an unexpected sight in the otherwise rugged landscape.

Orchid Beach derives its name from the Hyacinth orchid. Found growing all over the island, this pretty little plant survives on decaying plant matter, its distinctive pink-spotted flowers revealing its whereabouts in Spring.

## Getting there

From the south, drivers must travel up Seventy-Five Mile Beach before taking the inland track at the Champagne Pools. Much of this is one-way and the steep banks provide little opportunity for overtaking. Follow signs to Orchid Beach. The tide needs to be sufficiently low and conditions favourable for any beach driving so plan ahead and allow plenty of time. Also, the track from the Champagne Pools can be very soft and powdery at times.

If continuing north, there are a couple of tricky bypasses that require some skill (see Sandy Cape p 70). The journey across to Wathumba Creek on the west coast is also very scenic, as the track meanders through different vegetation types.

## Facilities

Known as the Orchid Beach Trading Post, the general store has most provisions, including eftpos, liquor and fuel (diesel, petrol and unleaded). In addition, it handles holiday home rental enquiries, basic vehicle and tyre repairs and provides a vehicle recovery service. Public toilets and drinking water are provided as well a post box and payphone. A grassed airstrip is situated in front of the store.

# Orchid Beach

*More sheltered than Seventy-Five Mile Beach, Orchid Beach is a very popular fishing spot, with picturesque deep green water and rolling waves.*

Campers can choose from a couple of beachside bush camping areas on top of the dunes - look for the beach turnoffs north of Waddy Point. Here, you have the advantage of being able to set up much closer to the water and on a more sheltered surf beach than Seventy-Five Mile Beach.

## What to look for

Although submerged, the wreck of the Marloo can sometimes be seen as a dark shadow less than half a kilometre out to sea. Try spotting it from the northern end of the airstrip. The Italian luxury liner, after which Marloo Bay was later named, took on water in 1914, coming to rest directly in front of where the Orchid Beach settlement is today. It's also worth having a look for marine wildlife. Dolphins sometimes travel in large pods, treating lucky onlookers to highly acrobatic displays.

Wild horses or 'brumbies' are often seen grazing on people's lawns. From the 1870's, during the island's logging days, Arabians were bred on a grazing lease near Orchid Beach. The wild horses that resulted, however, caused damage to the dunes and native vegetation and their grazing habits encouraged the spread of exotic pasture grasses. Today, following culling and relocation programs, only a very small herd remains.

## What we think

One of our favourite camping spots is around here…

In our view, many of the holiday houses don't reflect Fraser Island's 'wilderness' character. While they are likely to be perfect for many people who value their creature comforts and sea views, we have to admit we prefer accommodation that promotes a more natural setting (without the introduced garden plants).

# Northern Fraser Island

**Map reference**: G11 (Hema), D8 (Sunmap)

**Location**: 7 km from Champagne Pools, 9 km from Indian Head

**What's there?** Great beach fishing, small rocky headland, camping with good facilities

**Nearest to**: Champagne Pools, Indian Head, Orchid Beach

## Why go?

Beautiful deep green waters and rolling waves give the beaches north of Indian Head an irresistible charm. They are also a lot calmer than Seventy-Five Mile Beach. Add to that some of the island's best fishing and it's no wonder Waddy Point is so popular.

Thousands of anglers visit Fraser Island each year, especially from August - September during the tailor season. Waddy Point is the site of the famous annual Toyota Fishing Classic and a favourite spot for fishing enthusiasts all year round.

Many launch small boats from the beach (not recommended for the inexperienced) while others fish the deep gutters close to shore or off the rocky headland. The rocks and a nearby coral reef contribute to the diversity of fish species as well as the more sheltered conditions.

The Waddy Point headland is much smaller than Indian Head but both have the same volcanic origin. Comprised of rhyolite, it is one of the island's few outcrops of rock. It is not known how the area came to be called 'Waddy'; Aborigines referred to the point as 'Binngih'.

In 1836, Eliza Fraser and some of the crew of the shipwrecked Stirling Castle were forced to come ashore near Waddy Point. They were greeted by local Aborigines and put to work, performing tasks that were a normal part of Indigenous life of the island. Eliza was 'rescued' seven weeks later and went on to travel and make money recounting what she described as a cruel ordeal.

## Getting there

It isn't possible to drive on the beach around the rocky headlands - you'll need to take the inland track that starts just before Middle Rocks. The drive is very scenic with lots of bracken, wattles and banksias. At the Champagne Pools, the track splits and becomes one way for most of the journey to Waddy Point. The banks are quite steep, with little opportunity for overtaking and conditions can become soft and powdery when dry. On the return journey, the descent affords spectacular views of the ocean, headland and sandblow.

If driving on the beach at Waddy Point, watch for lagoons and pools of water.

## Facilities

The QPWS campground at Waddy Point offers a range of bush and beach camping options and has excellent facilities. Campsites can be booked in advance and this is recommended for Easter and other busy periods (details p11). Communal fire pits are provided but you'll need to bring your own clean firewood (see p11).

With its nice bush setting, the dingo-

# Waddy Point

*Made famous by the fishing competition that attracts thousands of anglers each year, Waddy Point is synonymous with great fishing, camping and ocean beaches.*

fenced upper campground has coin-operated hot showers, toilets and a few picnic tables. Nearby is a fully enclosed barbecue area with free communal gas barbecues, shelters, lights and ocean views. Drinking water, a pay phone and interpretive signage are located near the Waddy Point Ranger Base.

The lower campground offers beach camping and is popular amongst those with small boats (sites can be booked). Cold showers and toilets are provided.

## What to look for

Brumbies are often seen but keep your distance since wild horses can kick and bite. Lesser crested terns often congregate on the beach while birds of prey, like ospreys and sea eagles, may be spotted overhead. Dingoes, whales and large pods of dolphins may also be seen.

There are some great walks to help visitors take in the sights - the Waddy Point campground to headland walk (3·2 km return), Binngih Sandblow track (750 m return) and the walk to Champagne Pools (7 km return). A longer walk involves heading north along the beach to Ocean Lake when tides permit. While a vehicle track leads to the other side of Waddy Point, note that vehicles are only allowed on the first 100m or so of beach.

## What we think

The beach here is absolutely magnificent and while the campground is great, we usually opt for the privacy of the free-range beach camps just north.

As soon as you head north of the Champagne Pools, you experience a different, less touristy and more relaxed Fraser Island. You really feel like you are away from it all. Anyone into fishing and great beaches will love Waddy Point.

# Northern Fraser Island

**Map reference**: G11 (Hema), D8 (Sunmap)

**Location**: 2 km from Indian Head

**What's there?** Natural rock pools, small beach, ocean views

**Nearest to**: Indian Head, Orchid Beach, Waddy Point

## Why go?

Waves crashing over the seaward edge of a series of natural rock pools create a visual and swimming attraction at what have come to be known as the Champagne Pools or Aquarium.

As the pools fill with seawater, millions of tiny air bubbles rush to the surface like 'champagne bubbles', mixing and churning until the next wave strikes. Visitors enjoy watching the spectacle from the boardwalks above or taking a swim; an increasingly popular activity with younger visitors (especially backpackers). Sunbathers also make use of the small, secluded beach.

*One-way track heading south*

The Champagne Pools offer the only relatively safe saltwater swimming on the eastern side of the island. Elsewhere, sharks and rips make conditions potentially dangerous for swimming or surfing.

Spectacular ocean views and the chance to see a formation of significance to the island's traditional Indigenous owners are other reasons for visiting the rock pools. Aborigines once used such pools as natural fish traps, using spears and other techniques to capture fish stranded in the pools at low tide. Even today, a variety of fish often share the pools with swimmers.

## Getting there

The Champagne Pools are located just north of Indian Head. From the eastern beach, a short inland bypass leads to a small picturesque bay on the other side of the headland. Providing the tide is sufficiently low, drive along this beach, taking the next inland bypass to the Champagne Pools car park. It is not uncommon for vehicles to become bogged in the powdery, wind-blown sands along the beach and bypasses, especially near Indian Head (see Beach access points p 24).

## Facilities

Toilets and rubbish bins are located at the carpark. A boardwalk, dotted with interpretive signs, enables visitors to walk down to the pools. Note the last section requires climbing down over the rocks.

## What to look for

Birds of prey circling high overhead are a common and spectacular sight on this part of the island. Osprey, white-breasted sea eagles, and whistling and brahminy kites are also worth looking out for. From above, it is sometimes possible to spot sharks, dolphins, marine turtles or even humpback whales, depending on conditions and the time of year.

The rock pools offer the perfect place to swim or snorkel. Fish such as whiting, bream and sometimes tailor can be seen

# Champagne Pools

*At the island's famous Champagne Pools, cool ocean waters pour into the natural rock pools once used as fish traps by local Aborigines.*

and it is easy to imagine why Aborigines valued these pools as a source of food.

On land, larger mammals such as dingoes and wild horses known as brumbies may be seen, although cullling programs may have reduced the numbers of both.

## What we think

When conditions are right, the Champagne Pools offer an invigorating and salty alternative to swimming in the inland lakes and the surrounding ocean scenery is well worth a look.

If you're planning to make a special trip to the pools, check to make sure they're not filled with sand or fireweed (microscopic marine animals that sting and cause irritation - see p 65). To find out, try asking an island local or ranger.

Safety in the water is the responsibility of visitors and there are no life guards. Ocean currents are strong and unpredictable and the edges of the rock pools can be sharp and slippery. There is almost always someone who decides to walk along these precarious edges only to lose their footing and cut their feet. A first aid kit can be handy for cuts or marine stings. Freak waves and extreme weather also pose a risk. Observe the conditions on the day, use some common sense and you should have an enjoyable and safe experience.

*View towards Waddy Point*

# Northern Fraser Island

**Map reference**: H11 (Hema), E8 (Sunmap)

**Location**: 28.5 km from The Pinnacles, 33.5 km from Eli Creek

**What's there?** Large rocky outcrop, ocean beaches, 360 degree views, fishing, marine wildlife

**Nearest to**: Champagne Pools, Orchid Beach, Waddy Point

## Why go?

The impressive sight of a rocky headland protruding from an entirely sandy landscape makes Indian Head a unique landmark and popular tourist attraction.

Indian Head is the island's most easterly point. While many visitors are content to appreciate it from below, others walk to the top to enjoy the panoramic views and opportunity to spot wildlife.

*Pandanas growing on the northern face*

The name 'Indian Head' was given by Captain Cook in 1788. While on his famous Voyage of Discovery, he and his crew noticed a group of Aborigines who had gathered on the headland. Drawing on the language of the times, Cook decided to name the landmark after them.

Indian Head and Waddy Point are the only visible outcrops of an ancient hilly landscape that extends beneath the island, well hidden beneath the sand and lakes. The only true rock on the island, these outcrops of igneous rock (rhyolite) were originally created by volcanic activity during the upper Cretaceous / early Tertiary period, 50-80 million years ago.

Over time, sand traveling northwards is thought to have banked up against these rocky outcrops, the vegetation helping to trap it. Today, longshore currents bring sand from the south-east and strong waves continue to batter the island's eastern coast because of the steep and narrow profile of the Continental Shelf.

Much of the sand blanketing Fraser Island is said to have originated from the weathering of sandstones and granites in northern New South Wales and southern Queensland. Over time, rivers carried the sand out to sea and the sea currents transported it northwards. But sand may have also originated from as far as Antarctica, weathering of the Continental Shelf, underwater headlands and the sea floor - how much is not clear.

## Getting there

Indian Head can only be approached from the south by driving along Seventy Five Mile Beach or from Orchid Beach or Waddy Point just north, following a combination of beach and short inland tracks. Driving north towards Indian Head, the beach becomes impassable at most high tides. The inland bypass, just before the headland, is on a slight incline and is often very soft and powdery. Some momentum is needed to get through (see Beach access points p 24) but take care, particularly if bogged vehicles (and their distracted passengers) are present.

## Facilities

Parking space, payphone and rubbish bin facilities are provided. Note that Indian

# Indian Head

*As you drive up the beach, the haze from the surf slowly gives way to the sight of a striking, rocky headland - Indian Head.*

Head and the surrounding area is now permanently closed to camping.

## What to look for

Indian Head provides a unique vantage point for spotting animals that inhabit the air, sea and dunes. On a clear day, the climb to the top affords spectacular views in all directions and is one of the best places on the island to spot wildlife.

Reasonable fitness and agility are required to experience the 360-degree views. Steep walking tracks begin on both sides of the headland, winding their way to the top. Despite the numerous sheer drops, strong winds and treacherous surf below, there are no railings and the tracks are narrow and unmarked. It pays to take care and supervise children closely.

In the ocean waters below, it may be possible to spot manta rays, large schools of fish (eg tailor), sharks (eg tiger sharks, bronze whalers), turtles, dolphins or humpbacks. Bring your binoculars.

Birds of prey such as osprey, white-breasted sea eagles and whistling and brahminy kites are worth looking out for. Wild horses or 'brumbies' are sometimes sighted on the dunes, Tukkee Sandblow and in the old campground.

## What we think

The sight of the headland can be breathtaking and somehow puts the rest of the island into perspective. Take your time, this is a great place to take a break. On a clear day, you are almost certain to spot a turtle, ray, shark or some other sea creature from above.

Even if you're not able to climb to the top, we think the sight from below is well worth the trip. Those heading further north will find some of the island's most stunning ocean beaches just around the corner.

# Northern Fraser Island

**Map reference**: G8 (Hema), D6 (Sunmap)

**Location**: 16 km from Orchid Beach, 24 km from Indian Head

**What's there?** Sheltered beaches, fishing, stunning coastal scenery, large swamp and lagoon, good facilities

**Nearest to**: Awinya Creek, Orchid Beach, Waddy Point

## Why go?
Gentle breezes, calm waters and stunning coastal scenery attract visitors in pursuit of leisurely beach walks, boating, fishing and bird watching. For self-sufficient campers, Wathumba Creek offers the most northern established campground on the island's western shore.

Also a popular boat anchorage, the mangrove-lined inlet provides scenic views across sparkling blue-green waters towards the isthmus; a narrow stretch of white sand. The large movements of the tide constantly change the appearance of the inlet, with low tide providing a chance to explore the area on foot.

Upstream, the inlet widens to become one of the largest swamps on the island. Conservationists had to fight to protect this region. In 1973, a proposal to develop the sixty-five hectares at Wathumba Creek was successfully fought off by a local conservation group called the Fraser Island Defenders Organisation - a group that is still active today.

## Getting there
The only reliable route to Wathumba is from Orchid Beach heading westwards along Wathumba Road. Although there are many turns, the track is generally in good condition and is clearly signposted.

A short section of wooden planks leads out of Orchid Beach. From here on, the drive across the island is very scenic. Banksia woodlands with hundreds of grass trees blend into wetlands with sheoaks, smooth barked apple gums and ferns. Fallen sheoaks can sometimes block the track but are usually small and easily removed.

Further along, the drive takes on a 'lost world' feel as you skirt around the edge of a beautiful rainforest gully filled with lush cabbage tree palms and ancient cycads. Still to come are glimpses of the massive expanse of Wathumba Swamp to the right and beyond this, the creek mouth and camping area.

Do not attempt to use your vehicle to cross Wathumba Inlet. Conditions are extremely hazardous and vehicle access is prohibited north of this point. Many vehicles have been lost in foolish attempts to cross the inlet. It is possible to drive southwards, but the beach is narrow and conditions can be swampy and treacherous (see Awinya Creek p 90-91, Bowarrady Creek p 98-99).

*Woodlands, rainforest and wetlands make for a scenic drive.*

## Facilities
The day area and campground feature vehicle access, flushing toilets, show-

# Wathumba Creek

*Wathumba's natural beauty is a closely guarded secret amongst many visitors to Fraser's sheltered western shores.*

ers, picnic tables, and dingo-proof lockup units. About 20 grassed campsites are provided which operate on a first come first served basis. These can become busy during Easter and other holiday periods. You will need to bring your own drinking water since this is no longer provided.

## What to look for

From July to October, be on the lookout for humpback whales. The waters just offshore are one of their preferred resting and mating areas within Platypus Bay. Overhead, birds of prey such as the brahminy kite are a common site along with the campground's resident kookaburras and bar-shouldered doves.

Native mistletoes decorate the mangroves, their red fruits attracting mistletoe birds which, in turn, spread the sticky seeds to other branches. Below, armies of tiny soldier crabs march across the sand flats to feed at low tide, while larger mud crabs are at home in the mangroves. A variety of fish use the estuary for habitat and breeding grounds and there are plenty of mosquitoes and sand flies. Cover up because the insects can be severe.

By the beach is a rusty old 1930s shark boiler that was never used. According to the sign, it is all that remains of a shark oil factory and settlement at Wathumba.

## What we think

A very rewarding detour if you plan to explore the island's north. The beach and inlet offer a touch of paradise (if you don't mind the biting insects) and the track leading in is incredibly scenic. Wathumba Swamp is a breathtaking sight - wild, expansive and teeming with life. Friends of ours who've been visiting Fraser Island for years quietly admit this is their favourite destination. It's easy to see why.

# Central Fraser Island

From spectacular coloured sands to secluded lakes and breathtaking sandblows, the island's central region has plenty of great locations to explore.

The drive along Seventy-Five Mile Beach is a favourite with many island visitors, with Eli Creek, the Maheno and the Pinnacles providing popular stops.

Keen bushwalkers can enjoy some of the longer walks through the island's interior. Vehicle tracks are slow-going but the scenery varies from stunning rainforest to flowering heath and other forest types. Those who take the time to explore the west coast are usually rewarded with calm blue-green waters and idyllic beaches.

*Eli Creek*

*Woralie Creek*

*Yidney Scrub*

*Red Canyon*

*Sunrise, Seventy-Five Mile Beach*

WATHUMBA CREEK
ORCHID BEACH
CHAMPAGNE
GREAT SANDY
NATIONAL PARK
AWINYA CREEK
BOWARRADY CREEK
WORALIE CREEK
LAKE GNARANN
WHITE LAKE
LAKE BOWARRADY
RED CANYON
WUNGUL SANDBLOW
DUNDUBARA
LAKE ALLOM
BOOMERANG LAKES
COOMBOO LAKE
CATHEDRAL BEACH
MOON POINT
KNIFEBLADE SANDBLOW
THE PINNACLES
MAHENO WRECK
Big Woody Island
YIDNEY SCRUB
ELI CREEK
SOUT
LAKE GARAWONGERA
HAPPY VALLEY
PACIF
SANDY
VALLEY OF THE GIANTS
RAINBOW GORGE
KINGFISHER BAY
OCEA
er Heads
STRAIT
LAKE McKENZIE
CENTRAL STATION
STONETOOL SANDBLOW
LAKE WABBY
BASIN LAKE
PILE VALLEY
UNGOWA
LAKE BIRRABEEN
LAKE JENNINGS
EURONG
LAKE BENAROON

# Central Fraser Island

**Map reference**: I7 (Hema), E6 (Sunmap)

**Location**: 8 km (via beach) and 27.5 km (via inland) from Woralie Creek, 34 km from The Pinnacles

**What's there?** Beach camping, stunning sheltered waters and creek mouth, fishing, extensive wetland

**Nearest to**: Bowarrady Creek, Lake Gnarann, White Lake

## Why go?

Charmed by its small, white sandy beach, turquoise creek and calm ocean waters, those charmed by Awinya Creek return time and time again to this secluded spot.

Compared to the island's eastern beach, the scenery here is very different and the beach much more sheltered. Without the roar of the ocean and sound of passing vehicles, it's also a lot quieter.

Every visitor has their own way of soaking up the atmosphere of this unspoilt environment. Many enjoy camping, beach fishing, kayaking, walking, bird spotting, taking photos or just relaxing on their deck chairs. Despite the area's isolation, Awinya Creek is a favourite with young families. During busy periods, like Easter, they line their tents up along the beach, forming a happy, lively little community.

## Getting there

From Seventy-Five Mile Beach, take the turnoff onto the blue scenic track (Woralie Road), in the direction of Moon Point. Continue straight ahead, past the turnoff to Lake Allom, and take the signpost for 'Awinya Creek 16 km'. Nearer to Awinya Creek, keep an eye out for Lake Gnarann on the right. As you cross the island, the vegetation changes dramatically into low heath with flowering peas, swamp banksias, scribbly gums and grass trees. Look for the bright orange-coloured dodder vine and large vegetated dunes. There is always something in flower.

Some visitors choose to approach from Moon Point along Bullock Road but the journey can be slow-going and like any of the island's inland tracks, conditions can be quite soft, particularly in the dry. In favourable conditions, some choose to travel northwards along the beach at their own risk. On this side of the island, beach driving is permitted between Moon Point and Wathumba Creek, but drivers should be extremely cautious, particularly at creek crossings (see Driving on Fraser Island p 22-35).

Coongal Creek releases quite large volumes of water and has steep sides in places. The mouth of Bowarrady and other creeks can be hazardous due to patches of seaweed and quick sand. It is a good idea to get out and walk crossings first. Do not attempt to drive across Wathumba Inlet.

At the Awinya Creek mouth, some visitors drive across the creek at low tide (also at their own risk). If you decide to cross, it helps to have walked through first. The same goes when approaching from the inland track - many people cross the creek in order to reach the beach on the other side to fish or set up camp. Look for the entry and exit tracks from other vehicles. The creek can be quite deep

# Awinya Creek

*On a sunny day, changing tides and depths bring out captivating shades of blue, green and gold near the mouth of Awinya Creek.*

and fast-flowing so a snorkel (especially for vehicles with low air intake) and some momentum to create a bow wave is advantageous. Water may splash onto your bonnet on entry.

These two crossing points should be treated with extreme caution and at high tide or in unfavourable conditions, it may not be possible to cross.

## Facilities

Although free-range beach camping is permitted, there are no facilities so visitors need to be entirely self-sufficient. The nearest settlement (by vehicle) is Frasers at Cathedral Beach on the other side of the island.

## What to look for

There is plenty to see and explore along the creek and beach as well as in the wetland and mangroves. Near the creek mouth, a pair of white-bellied sea-eagles may be seen nesting in the prominent exposed gum tree on top of the dune. On the beach, pied oyercatchers can often be seen; always in couples since they pair up for life. Their 'pee-pee-pee-pee' call is a familiar beach sound.

Crabs, fish and a variety of birds inhabit the estuary as well as biting insects, while dolphins and humpback whales are sometimes seen offshore. Kayaking is popular and keen walkers can head south along the beach at low tide to Bowarrady Creek (just over 5 km one way).

## What we think

No matter what the weather, we are always captivated by the unique beauty of Awinya Creek. It is literally brimming with paradise-like qualities, making it the perfect place to relax - if you don't mind the isolation and biting insects. Plenty of time, a comfortable deck chair and a good book are a must.

# Central Fraser Island

**Map reference**: J7 (Hema), F6 (Sunmap)

**Location**: 8.5 km from Awinya Creek, 18 km from Lake Allom

**What's there?** Small freshwater lake

**Nearest to**: Awinya Creek, White Lake, Lake Bowarrady

## Why go?

Located on the way to Awinya Creek, this small freshwater lake receives few visitors but can be a source of great interest to keen naturalists and photographers.

Although about a quarter the size of Lake McKenzie, Lake Gnarann is not a tempting option for swimmers. Prolonged dry spells can cause the lake to shrink considerably, leaving scattered pools of water and an unusual layer of charcoal-coloured organic matter. This provides a stark and interesting contract against the white sandy beaches.

*Carnivorous sundew*

Visitors to the area enjoy some fabulous views as they continue on towards Awinya Creek. Shortly after Lake Gnarann, the track leads up and along the top of a ridge, affording great views off either side. It then loops around, making it possible to spot the lake from above.

## Getting there

Lake Gnarann is accessible along Bowarrady Road and can be approached from Awinya Creek or Woralie Creek, the inland tracks just south, or Seventy-Five Mile Beach to the east.

From the eastern beach, take the turnoff just south of The Pinnacles, following the Woralie Track for 13.5 km before taking the turnoff for Awinya Creek. After the hairpin turn, Lake Gnarann comes into view on your right hand side. (The lake has a very reedy appearance). Keep an eye out for the very small parking bay and sign located directly in front of the lake. The inland tracks can be soft and slow-going but enjoy little traffic and great scenery.

## Facilities

There are no facilities located at the lake or in the vicinity, the nearest settlement (by vehicle) being Frasers at Cathedral Beach on the island's east coast.

## What to look for

The lake is surrounded by a wonderful diversity of plants, including plenty of paperbarks, grass trees, bracken ferns, banksias and flowering shrubs.

Carnivorous plants are easy to spot along the edge of the lake. Look for the bright red colours of the sundew - a small, unusual plant that exudes a sticky substance to attract and trap insects. Fraser Island is home to at least five different types of sundew. *Drosera* species, as they are scientifically known, are unique

# Lake Gnarann

*Banksias are a common sight in the forests surrounding Lake Gnarann.*

enough to have earned a classification of their own - the Droseraceae family.

Freshwater turtles can occasionally be seen and when the lake has been topped up with rain, it is possible to spot visiting water birds. A variety of woodland birds inhabit the surrounding forest, including the striking emerald ground-dove.

During their breeding seasons, native 'acid frogs', like the tiny Cooloola sedgefrog and wallum rocketfrog, may be heard. Attracted to their calls, snakes such as keelbacks and pythons prey on frogs, usually by night, but these particular snakes are not dangerous to humans.

On a sunny day, you may be lucky enough to spot a snake crossing the track. Wait for the snake to pass and keep well clear as it could be one of the island's highly venomous species. Keelbacks and venomous brown snakes, for example, can be very difficult to tell apart.

## What we think

Because it's so close to the road, Lake Gnarann is definitely worth a quick look if you're on your way to Awinya Creek. It's an interesting spot to have a little explore and enjoy the peace and quiet.

*This exposed lake provides an interesting stop on the way to Awinya Creek*

# Central Fraser Island

**Map reference**: J8 (Hema), F6 (Sunmap)

**Location**: 2 km from Lake Gnarann, 10.5 km from Awinya Creek, 18 km from Lake Allom

**What's there?** Small freshwater lake, a sense of isolation

**Nearest to**: Awinya Creek, Lake Bowarrady, Lake Gnarann

## Why go?

White Lake is a relatively large, secluded lake, surrounded by rushes, paperbarks and vegetated dunes.

Although not appealing as a swimming lake, the water's edge teems with an abundance of fascinating plantlife.

*Old paperbarks (Melaleuca sp.) take on gnarled, twisted forms.*

Since few people take the time to visit, bushwalkers, nature enthusiasts and photographers can enjoy exploring the shoreline in relative quiet and solitude.

## Getting there

Vehicle access is along Bowarrady Road, following signs to Awinya Creek. The drive is very scenic but can be soft and slow-going, particularly in very dry conditions. Keep an eye out for the small signposted parking bay located on the edge of the hairpin bend - there is just enough room for one car to pull in off the track. The lake is then a 1 km walk, mostly along an open track that used to provide vehicle access to Lake Bowarrady (heading south-east). The short walking trail down to the lake, however, is very steep and overgrown. Located on the left, the beginning of this path is marked by a small sign that may be obscured by vegetation.

## Facilities

There are no facilities located here or anywhere in the vicinity.

## What to look for

A massive scribbly gum is one of the first things to grab your attention on the interesting walk to the lake. The numerous hollows presumably house a variety of animals and insects. There's also plenty of grass trees, midyim berries and a beautiful pink smooth-barked apple gum strangely supported by three trunks.

It soon becomes obvious that the local plants are trying to reclaim the old vehicle track, the multitude of seedlings testimony to the bush's amazing capacity to regenerate itself. Numerous holes, scratchings and diggings suggest plenty of animal activity and one or more of the island's 300 ant species excavates sizable nests along the track.

As soon as you begin to catch glimpses

# White Lake

*Beautiful yet bleak, White Lake is an interesting place to explore and reflect on nature.*

of the lake to the left, keep your eyes peeled for the start of the walking trail down to the lake. Reasonable fitness and mobility are needed in order to step over large logs and maintain your footing on the steep, slippery path. At the base, it's a good idea to draw arrows in the sand so that you can relocate the start of the track. Without some kind of marker, it may prove impossible to find.

Try your luck spotting dragonflies and birds such as plovers, musk ducks and even wedge-tailed eagles. Although more elusive, freshwater turtles can also be seen along with the often colourful soft-spined sunfish.

Along the beach, there are plenty of signs of limited visitation by people. Seedlings, representing many different plant species, including banksias, paperbarks and sheoaks, pop up in every available space. Sedges and reeds flourish in the shallows and the lack of swimmers has left the layer of organic material lining the lake bed undisturbed. It's also possible to see distinct zones of vegetation running parallel to the water, beginning with rushes in the shallows, a mix of pioneer species on the pale pink-coloured beach and clumps of sedges and rushes on the white sand behind. These are followed by bands of flowering heath, paperbarks, shrubs then finally woodland.

## What we think

Each of the island's lesser-known lakes is interesting and uniquely different. But although 'we' might find them fascinating, we don't expect everyone else to. One of the great things about White Lake is that you'll almost certainly be the only people there. The lake's bleak appearance, subtle colouring and pristine qualities are sure to satisfy the inquiring mind of any enthusiastic nature lover.

# Central Fraser Island

**Map reference**: K8 (Hema), F6 (Sunmap)

**Location**: 11 km walk from Dundubara (via Bowarrady Trail), 15 km from Awinya Creek, 31 km from The Pinnacles

**What's there?** Secluded freshwater lake, rainforest, walking track, walkers' camp, turtles

**Nearest to**: White Lake, Lake Gnarann, Dundubara

## Why go?

Surrounded by magnificent rainforest, Lake Bowarrady is a large, freshwater lake that attracts visitors who enjoy bushwalking, bush camping, swimming and nature appreciation.

The lake was once a popular tourist destination, accessible via a long circuit drive. Turtles were one of the main attractions and the years of disturbance associated with people swimming forged a small beach. The recent move to restrict this track to walkers and management vehicles has restored peace and tranquillity to this secluded spot.

Nearby Mount Bowarrady, which stands at 244 metres above sea level, is the highest point on the island.

## Getting there

Walking is the only means of accessing the lake and because there are several tracks to choose from, those interested should obtain information from QPWS (see p 14 -16). A popular route is along the Bowarrady Trail from Dundubara on the east coast (11 km one way) while a shorter option is along the management track that passes the southern side of White Lake to the northwest. The start of the latter is located on Bowarrady Road, by following signs to Awinya Creek. From your vehicle, you can take in some amazing views across the landscape as the track meanders up to the top of a hill. Driving can be slow-going, however, especially if conditions are dry and the sand is soft. Park in the small signposted parking bay located on the edge of the hairpin bend - there is just enough room for one car to pull in off the track.

## Facilities

Apart from the hikers' camp, which consists of a large clearing, there are no facilities. Walkers should therefore be entirely self-sufficient.

## What to look for

The walks to Lake Bowarrady are very scenic, passing through open eucalypt woodlands into wallum heathlands that produce a profusion of flowers all year round. Because of the exposed conditions, take plenty of water and avoid hot weather. Walking is a great way to appreciate the subtle beauty of the wallum, with the most colourful displays during Spring.

Keep an eye out for the brightly orange

*Walk through a pocket of rainforest*

# Lake Bowarrady

*Turtle spotting, lush rainforest and a refreshing dip await those who undertake the long walk to Lake Bowarrady.*

coloured dodder vine, white flowering shrubs like wedding bush and woombye, and 'furry' green foliage of foxtail ferns. You will also spot plenty of animal holes and diggings. The track from White Lake follows stunning sedgelands full of 'islands' of *Gahnia*. Nearer the lake, the forest canopy begins to close over, as tall woodlands replace the wallum, grading into impressive stands of cool, dark rainforest. From the walkers' camp / old carpark it is a further 2km to the lake.

During its busier days, the lake's resident turtles were notoriously tame. Nowadays, they are much more elusive and visitors are encouraged not to feed them. With a little patience and a sharp eye, however, you should be able to spot some.

Known as Kreft's river turtles, these creatures are a kind of short-necked turtle belonging to the Chelidae family. Its members, which include the Brisbane short-necked turtle, typically have webbed, clawed feet and the ability to tuck their heads in sideways. Scientists suspect that the island's turtles might in fact represent a new species, quite different to Kreft's river turtles that inhabit the mainland. Most of the island's lakes are thought to contain turtles. Scavenging sand goannas sometimes dig up and feast on their eggs. The lake is also home to firetail gudgeons; a small freshwater fish that has been bred in aquaria.

## What we think

Bushwalkers will love the beauty and privacy of this once popular lake, plus the varied scenery along the way. Although the tracks and trail are relatively easy to walk on, walkers need to be moderately fit. Once at the lake, you're a long way from help and may not encounter any other walkers, even in peak holiday periods. For some, the serenity this affords may be part of the appeal.

# Central Fraser Island

**Map reference**: J6 (Hema), F5 Sunmap)

**Location**: 3 km from Woralie Creek, 5.5 km from Awinya Creek (both via beach)

**What's there?** Tea-coloured creek & water hole, beach, bush camping

**Nearest to**: Awinya Creek, Lake Gnarran, Woralie Creek

## Why go?

The deep, tea-stained waters of Bowarrady Creek and nearby coloured sands make the mouth an interesting stop for walkers, campers and day visitors.

Large volumes of gold and brown coloured water travel over a bed of coffee rock and soft organic 'slush' before spilling into the sea. At nearby Arch Cliff and further south, impressive sand cliffs exhibit spectacular shades of brown, charcoal, coffee and gold beneath the fragile layer of white quartz sand. Below, the appearance of the beach is always changing. Weed washes up en masse and outcrops of coffee rock are buried and exhumed.

*Colourful sand cliffs, seaweed in front*

Most of the time, the sea is sheltered from the wind, so much so that from July-October, the blows and other surface antics of humpback whales may be spotted some distance away. On calm, sunny days, the ocean takes on stunning shades of green and blue.

Back from the beach, where the creek is quite wide and deep, a makeshift swing suggests occasional bouts of popularity as a swimming hole.

## Getting there

The only access to the Bowarrady Creek mouth is by walking northwestwards for 4.5 kilometres along the management section of Bowarrady Road or walking or driving along the beach. Although beach driving is permitted between Moon Point and Wathumba Creek, conditions can be treacherous, especially at high tide (see Awinya Creek p 90-91 and Woralie Creek p100-101). Drivers should walk the mouth of Bowarrady Creek first - hidden hazards include buried coffee rock, seaweed and quick sand. Use extreme caution and only cross creeks if conditions allow.

## Facilities

There are no facilities, just two small clearings located either side of the creek. Campers should be fully self-sufficient and set up at least 50 metres from the creek. While it is recommended you bring your own drinking water, some people choose to boil or treat creek water collected upstream.

## What to look for

A paddle up the creek in a canoe or kayak is a great way to spot wildlife. At low tide, fiddler crabs emerge from many of the intertidal sandbanks here and elsewhere along the island's west coast.

Male fiddlers are easy to spot; each has one oversized orange-coloured claw used to fight other males and, by waving, attract the attention of females. But they

# Bowarrady Creek

*Shaded by paperbark trees, Bowarrady Creek provides the vital link between fresh and salt water for a diversity of local plant and animal life.*

can only eat using their smaller claw, and so must devote more time to feeding than the comparatively dexterous females. As the tide comes in, each crab returns to its burrow, sealing it tight with a mud plug and bubble of air.

Mudskippers are another common burrow dweller, also active at low tide. An unusual type of fish, they leave the water in pursuit of prey. Thanks to their specially modified pectoral fins, mudskippers can propel themselves along with surprising speed and agility. They can breathe both in and out of the water as well using their gills and other parts of their bodies. When danger approaches, they quickly retreat into their burrows.

Crustacea, fish, worms and insects are a food source for a diverse range of birds and other wildlife. Because of the diversity of habitat types, this is a great place for bird watching enthusiasts. Waders can often be spotted, including eastern curlews, whimbrels and pied oystercatchers. Kites, sea eagles and osprey hunt from above and visiting brown boobies impress lucky onlookers with their spectacular dives. Australian pelicans are often seen at rest or scooping up fish with their large bills in the shallows. By congregating in small flocks and swimming in formation, these giant birds are skillful fish herders.

Like most places on the western shore, biting midges and mosquitoes can be a problem so come prepared.

## What we think

Although not as picturesque as Awinya, but sharing many of the natural characteristics of Woralie Creek, Bowarrady Creek is a very pleasant spot for those wanting to relax and enjoy nature. When the other creeks are busy, this is a quiet west beach camping alternative for walkers and nature enthusiasts - providing you are fully self-sufficient.

# Central Fraser Island

**Map reference**: J6 (Hema), F5 (Sunmap)

**Location**: 12.5 km from Lake Allom, 21.5 km from The Pinnacles

**What's there?** Tea-coloured creek, beach, small clearing, bush camping

**Nearest to**: Awinya Creek, Bowarrady Creek, Lake Gnarran

## Why go?
At the mouth of Woralie Creek, stunning brown and gold-coloured water cascades over coffee rock before emptying onto the beach and into Hervey Bay.

Campers, walkers, photographers and nature lovers enjoy the unique and interesting beach scenery of the creek mouth. The predominant south-easterly winds create calm, sheltered conditions. When the sun is out, the sea turns a stunning shade of turquoise making this spot very picturesque.

*Brown turns to gold against the creek's white sandy banks*

## Getting there
The creek mouth is located at the end of the Woralie Track on the western side of the island. From the eastern beach, this track starts just south of The Pinnacles, taking in some stunning scenery and a host of plant communities as it crosses the island. Nearer to the west coast, look for the dried up lake bed on the left.

Both this and the inland route from Moon Point can be slow-going, with sometimes very soft conditions and deep tyre ruts. Bullock Road often splits into two tracks then re-forms. When conditions permit, a few people choose to drive along the beach from the north or south at their own risk. While beach driving is permitted between Moon Point and Wathumba Creek, drivers should be extremely cautious, particularly at creek crossings (see Driving on Fraser Island p 22-35). These creeks may be impassable, even at low tide, due to patches of buried weed and quicksand. It is a good idea to get out and walk crossings first.

While the drive along the beach to Bowarrady Creek features some stunning coloured sand cliffs, seeping freshwater and protruding coffee rock, hidden under deceptively-firm 'fields' of weed, can potentially make the going treacherous, especially at high tide.

## Facilities
There are no facilities, apart from a small, partially shaded clearing near the beach that can accommodate a small number of bush campers. Campers should be fully self-sufficient. Drinking water is not provided, nor is camping permitted within 50 metres of the creek.

## What to look for
The creek's tea-coloured water provides fantastic photographic opportunities, especially when the sun shines overhead. It derives its striking colour from organic acids leached out of decomposing vegetation. Before flowing out to sea, this

# Woralie Creek

*One of the island's larger freshwater creeks, Woralie Creek carries tannin-stained water for several kilometres before pouring out to sea.*

'black' water passes through white quartz sands. Unlike yellow-brown sands, there is nothing for the acids to react with so the acids remain in the water, giving the creek its characteristic tea colour.

Native gums and Cypress pines encircle the small clearing. See if you can recognise the toothed wattle, *Acacia flavescens*, with its extra-large leaves. Over by the creek, the gnarled, twisted forms of paperbarks are at home in the boggy, anaerobic soils. Look for their flaking paper-like bark and black and white trunks from which the tree's Latin name, *Melaleuca* (meaning 'black white') is derived.

Bird enthusiasts can enjoy spotting a variety of woodland, mangrove and beach birds, including rainbow bee-eaters, azure kingfishers, pied oystercatchers and pelicans. If you're lucky, you might hear the beautiful call of a mangrove honeyeater. When breeding, these accomplished vocalists gather together to sing and fight. Astonishingly, they build cup-shaped nests in which cuckoos sometimes lay their eggs. The honeyeaters may then be tricked into raising the cuckoos' chicks.

While most birds are active early in the morning, nocturnal hunters, like owls and tawny frogmouths, prey on insects and small nocturnal mammals under the cover of darkness.

## What we think

Most of the time, expect to get this unspoiled, photogenic spot to yourself. Occasionally, people may drive past or stop to fish off the beach but passing boats are a more common sight. While the solitude and scenery will hold special appeal for a few, we prefer Awinya or Wathumba as kayaking and camping destinations. At all of these places, be sure to guard against biting insects.

# Central Fraser Island

**Map reference**: K9 (Hema), F7 (Sunmap)

**Location**: 3 km from Dundubara, 13 km from The Pinnacles

**What's there?** Red-coloured sands, ocean beach, fishing

**Nearest to**: Cathedral Beach, Dundubara, Wungul Sand Blow

## Why go?

Red Canyon is one of the island's most impressive coloured sand formations. As its name implies, the formation's intricate shapes are dominated by striking shades of reds and orange. These have been carved by out the wind and rain.

The exposed cliff face provides a unique opportunity for a close-up look. While coloured sands stretch for many kilometres along Cathedral Beach, large sections are hidden beneath a protective blanket of vegetation. These are said to be some of the most spectacular coloured sand formations in the Cooloola sand mass.

*Coloured sand dunes extend for miles*

Without coloured sands, the sand masses would not exist. Many of the minerals that coat the sand grains, and give rise to the varying colours, are crucial for plant growth. Amongst these are iron and aluminium and, to a lesser extent, potassium, phosphorus calcium and sulphur.

With the help of nitrogen-fixing species, plant communities establish and as the process of succession continues, great sand masses like Fraser Island are eventually able to form. The white sand that dominates many of the lakes' edges and island's older west coast is deficient in plant nutrients - hence these areas' more stunted and scrubby vegetation types.

For the island's traditional owners, the Butchulla, the coloured sands are sites of significance and cultural identity. Their account of how the colours were created is a story of love and revenge, involving a young girl and Wiberigan (the rainbow). Other coloured sands can be seen at The Pinnacles, Rainbow Gorge and elsewhere along Cathedral Beach.

## Getting there

Red Canyon is located a few kilometres north of Dundubara along Seventy-Five Mile Beach. Parking is high on the beach, in front of the canyon. As a courtesy, you may wish to park a little off to the side, so as not to spoil people's photos.

Aim to drive on the beach at or as near as possible to low tide and take care when crossing creeks. A short, sandy path leads into the canyon.

## Facilities

There are no facilities provided, except an information board.

## What to look for

Not unlike the red-coloured earth of central Australia, the deeper reds and oranges of the canyon reflect concentrations of iron oxides. The lighter shades, on the other hand, have been leached of the rusty-coloured compound.

# Red Canyon

*Red Canyon offers the ideal stop along stunning Cathedral Beach.*

Red Canyon is usually at its best on a clear day, early in the morning when the rising sun lights up the dune face.

Don't be tempted to take a souvenir sand sample - this can cause damage and there are hefty penalties. For the same reason, keep to the designated path, well away from the cliff face.

Dingoes are sometimes spotted on the beach, often scavenging for dead birds or fishy morsels left behind by the receding tide. Occasionally, visitors witness a pack traveling along the beach. (See p 44-47 for more information on dingoes).

The beach is also a great place to spot birds. Keep an eye out for pied oyster-catchers (always present in pairs) or large flocks of crested terns.

Sharks may be seen fishing the gutters close to shore and much further out, you may be lucky enough to spot the 'blow' or 'round-out' of a humpback whale as it makes its long journey to or from Antarctica.

## What we think

Red Canyon is definitely worth a look if you're heading up the beach or camping at Dundubara. And while the red sand looks great in any lighting, keen photographers will need to get there just after sunrise for the best pic.

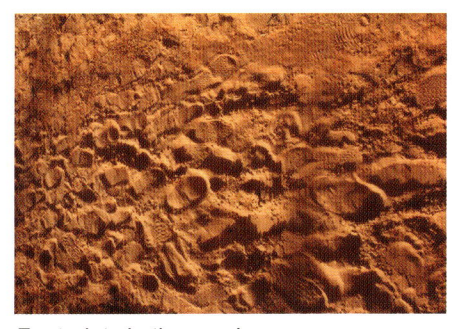

*Footprints in the sand*

# Central Fraser Island

**Map reference**: K9 (Hema), F7 (Sunmap)

**Location**: 2 km (return) from Dundubara

**What's there?** Walk onto sandblow with panoramic views, Dundubara campground

**Nearest to**: Dundubara, Lake Bowarrady, Red Canyon

## Why go?

Visitors who don't mind a little sand in their shoes can enjoy the spectacular interplay between sand, vegetation and the elements at Wungul Sandblow.

A natural feature of the island's eastern coast, sandblows have dramatically shaped and reworked the landscape over many thousands of years. They begin when wind funnels through a gap in the beach vegetation, blowing sand inland until enough momentum and mass is created to form a moving dune front.

*Red and white posts mark the way*

On top of Wungul Sandblow, the wind etches out wave-like sand ridges. This is where many of the postcard images depicting golden sand dunes against crisp blue skies have been taken.

By trekking to the top of these ridges, walkers can take in the 360 degree views that sweep towards the ocean and outer edges of the massive sand expanse. The interface between sand and vegetation is always changing as the dune relentlessly forges inland.

## Getting there

Access by vehicle is only possible by taking the Dundubara turnoff, situated 6.5 km north of Cathedral Beach along Seventy-Five Mile Beach. Aim to drive as near as possible to low tide, applying particular caution at tricky places like Eli Creek.

The sandblow, which can only be viewed on foot via a short walk (2 km return) and 5.5 km circuit walk, is reached from the western end of Dundubara campground. It requires reasonable fitness since the sand can be quite dry and soft and there are lots of ups and downs.

Circuit walkers must first climb up onto the sandblow and over numerous sand ridges. Take plenty of water - the circuit can easily take more than two hours. Hot weather should be avoided because of the exposed conditions. A little rain, on the other hand, hardens the sand and can make the going a little easier. Walking westwards over the dunes, follow the markers to a return track (marked by a red and white pole) which loops back through the forest. You can either return to the campground (via the firebreak link) or continue on towards the beach.

A much longer walk leads to Lake Bowarrady along the Bowaraddy Trail (11km one way). The vegetation changes dramatically, from dry open forest to lush rainforest nearer the lake. Walkers should allow a full day for the return journey and make appropriate preparations. Again, hot weather is best avoided. For any of the walks, it is recommended that you obtain information and advice from QPWS and refer to maps.

# Wungul Sandblow

*Stretching from the beach towards the island's vegetated interior, the desolate expanse of moving sand affords spectacular views in all directions.*

## Facilities

While there are no facilities on the sandblow itself, Dundubara has a QPWS office / visitor information centre and some basic facilities such as parking, drinking water and a payphone.

## What to look for

Despite the incredibly harsh conditions, it's possible to spot signs of plant and animal life on the sand blow. Lizards - and sometimes snakes and dingoes - leave their tracks behind while birds of prey hover above in the hope of an easy meal.

Although perhaps not as obvious at first, the trees and shrubs surviving along the edges of the sandblow show remarkable resilience and diversity of form. Sand buildup and erosion throw up many challenges for these adaptable species. Many of the plants are buried only to later re-sprout and reclaim their territory. At other times, they stand no chance against the sheer bulk of the dune and are smothered to death. When conditions later change, some are exhumed from their sandy tombs. Away from the leading edge, pioneer species like casuarinas colonise the dune, returning stability and nutrients to the sand and shelter from the salt-laden winds. In this way, they pave the way for secondary species and the process of forest succession begins again.

## What we think

The walk - or rather 'stagger' - out onto the sandblow will capture the imagination of anyone who loves breathtaking views and has plenty of energy to burn. It's also a great place to snap some great photos and take in the sheer size and energy of the world's largest sand island.

# Central Fraser Island

**Map reference**: K9 (Hema), F7 (Sunmap)

**Location**: 10 km from The Pinnacles

**What's there?** Extensive campground with basic facilities, beach fishing, walking tracks, sandblow

**Nearest to**: Cathedral Beach, Red Canyon, Wungul Sandblow

## Why go?

With its abundance of shady, grassed campsites and pleasant holiday atmosphere, this spot is a popular choice with families.

Conveniently located on the island's east coast, Dundubara has plenty to keep campers and day visitors entertained.

Many enjoy fishing off the beach or just relaxing in the shade of the paperbarks. Those with more energy to burn can take in the view from the top of Wungul Sandblow, enjoy the long walk to Lake Bowarrady (22 km return), or take the short drive down the beach to cool off in Eli Creek - a favourite with chiidren.

*Information and well-lit kitchen facilities*

## Getting there

Dundubara can only be accessed directly from Seventy-Five Mile Beach. (The track from the island's west is now restricted to management vehicles and walkers). Aim to drive along the beach at or as near as possible to low tide. Take the signposted turnoff to Dundubara. The ranger station and campground is located just a short distance back from the beach.

## Facilities

The QPWS Information Centre provides a range of interesting facts and updates relating to the island. Day visitors can use the carpark and adjacent grassed area, which has several picnic tables and a solar-powered barbecue.

Other facilities include rubbish bins, a payphone, post box and drinking water. Information about the walking trail to Lake Bowarrady and the Wungul Sandblow circuit walk should be sourced from the ranger office (when open) or other QPWS offices (see p 21, 14 - 16).

The campground, which is managed by QPWS, features more than 90 campsites, lots of picnic tables and a large undercover kitchen with numerous sinks for washing up. Other facilities include coin-operated hot showers, flushing toilets, interpretive signage, communal gas barbecues and drinking water. Many of the sites are grassed and shady. Fish cleaning facilities are located on the beach, just 1.3 km north.

Campsites can be booked in advance (see p 11 for details). This is recommended for Easter and other busy periods. At such times, the campground may be fully booked so check with rangers before setting up.

## What to look for

Dingoes may be seen on the beach or

# Dundubara

*Just back from popular Seventy-Five Mile Beach, Dundubara's well-appointed campground is a favourite with families and groups.*

around the campground, so take heed of the advice given by rangers. Campers and walkers can enjoy a great variety of birdlife, from small wrens and honeyeaters to the larger kookaburras, crows and birds of prey.

Fishing is a popular pastime for many campers, with the ocean beach producing great catches at certain times of the year. The best time to catch tailor, for example, is August to September. The coloured sand formations exposed along the beach create a very pleasant backdrop.

The information provided by QPWS makes for interesting reading, as does the commemorative stone located next to the ranger office. The stone is dedicated to Ronald George Walk. The first ranger to take up residence and work on Fraser Island, 'Ranger Ron' served on the island from 1977 until his death in 1995.

The short walk up onto Wungul Sandblow provides spectacular views across the dunes towards the ocean and unique opportunities for taking photos (see Wungul Sandblow p 104 - 105).

## What we think

It's easy to see why Dundubara's campground is so popular with families and groups - there is plenty to do in the vicinity and the forest of paperbarks and absence of shops and infrastructure give the feeling of bush camping without having to 'really' rough it.

In other words, this is a great choice for campers who want to get back to nature without having to dig their own toilet. In peak holiday periods, it's probably a little too high density for us, but others no doubt love the lively atmosphere.

# Central Fraser Island

**Map reference**: L9 (Hema), G7 (Sunmap)

**Location**: 4 km from The Pinnacles, 9 km from Eli Creek, 13 km from Lake Allom

**What's there?** Seaside resort, camping, beach fishing, sand blows, coloured sands

**Nearest to**: Dundubara, Maheno Wreck, The Pinnacles

## Why go?

Located just back from a beach famous for its fishing and coloured sand dunes, this privately-owned resort serves as a popular base for exploring the area.

*Shady campsites*

The resort is named after 'The Cathedrals' - an 18 kilometre stretch of beach said to have some of the most spectacular coloured sand formations in the Cooloola sand mass.

Exposed by erosion, coloured sands like these are in fact thought to be buried under much of the island's surface.

## Getting there

The resort 'Frasers at Cathedral Beach' is situated on Seventy-Five Mile Beach just north of The Pinnacles. The only access is along this beach, preferably at or as near as possible to low tide. Look for the signposted turnoff - the resort and campground are just 400 metres back from the beach.

## Facilities

The resort's general store has a range of basic provisions, including eftpos, takeaway food, fuel, ice, souvenirs and liquor. A picnic area, car park, public toilets, post box and pay phone are provided.

Accommodation includes two and three bedroom cabins, which at times are fully booked, so it is best to book ahead. The campground has plenty of grassy campsites (including powered sites) situated under large shady trees. It is possible to book these in advance. (Note camping is not covered by the QPWS camping fee). Facilities include hot showers, toilets, drinking water and laundry facilities.

## What to look for

There's plenty to see and do in the area. Fishing is popular, with bream, tailor and whiting commonly caught off the beach. Swimming in the ocean is not advisable, however, due to sharks and dangerous rips. Plus there are no lifeguards.

Walking trails are located in the vicinity, or you can make the 6 km trek along the beach to the Maheno shipwreck.

Dingoes are sometimes seen on the beach and in the campground (see p 44-47 for more information). It's also worth keeping an eye out for dolphins and other marine life. From July to October, you have a good chance of spotting passing humpback whales (although you may need a pair of binoculars). The birdlife is also excellent, with a variety of species frequently seen around the resort. Up above, you may be lucky enough to spot a large bird of prey, like a sea eagle.

Note that planes take off and land along

# Cathedral Beach

*Tucked in behind kilometres of coloured sands, Frasers at Cathedral Beach serves up a relaxed atmosphere and shady bush setting.*

the beach and drivers are obliged to give way. The scenic flights can be very worthwhile, giving passengers a completely different perspective of the island, its sea of vegetation, track networks and abundance of lakes. Helicopter flights can also be booked from here.

On a clear day, early morning is usually the best time to view and photograph the coloured sand formations along Cathedral Beach. The colours can also look more vivid after rain. Iron oxides leach down through the layers, creating varying shades of red and orange. Spectacular contrasts are added by streaks of white sand and black organic material that has worked into the layers and been sculpted by wind and rain.

Take pictures instead of samples and keep well away from the base of cliffs and dunes. Even small disturbances can lead to accelerated damage and landslides can happen silently and without warning.

## What we think

With so many great attractions close by, this is an excellent choice for visitors wanting to see and do lots in a limited space of time. Eli Creek, The Pinnacles, sand blows, the Maheno and Lake Allom are all within twelve kilometres. Add to that some great walks and fishing and you can't go wrong.

Thanks to its shady bush setting, this pleasant spot shares many of the 'back to nature' qualities of Dundubara, only with some extra conveniences like the shop and accommodation options.

# Central Fraser Island

**Map reference**: L8 (Hema), G6 (Sunmap)

**Location**: 8.5 km from The Pinnacles, 14 km from Eli Creek

**What's there?** Small freshwater lake, scenic perimeter walk, freshwater turtles, beautiful forest

**Nearest to**: Knifeblade Sandblow, Maheno Wreck, The Pinnacles

## Why go?

One of Fraser Island's more secluded perched lakes, Lake Allom is a quiet, pretty destination that attracts families and nature lovers interested in its resident turtles and peaceful surrounds.

As well as turtle spotting, visitors to the lake enjoy swimming, picnicking, bird-watching and taking the scenic forty minute walk around the lake. Families find this area appealing, especially those with children wanting to swim and observe wildlife.

## Getting there

The easiest access to Lake Allom is from the eastern beach, taking the Woralie Road beach turnoff immediately south of The Pinnacles. Once off the beach, simply continue along Northern Road, turning right at the signpost for Lake Allom.

Changes in vegetation make this a very interesting and scenic drive. You'll see plenty of brushbox and satinay trees and pass through stands of subtropical rainforest as beautiful as those at Central Station.

It is also possible to approach from the opposite direction, along Northern Road, although this is a much slower, more time-consuming option.

The track into Lake Allom is generally okay but quite narrow and often rough in places, so be patient and take your time.

## Facilities

Located a short distance from the lake are picnic tables, toilets (wheelchair access) and a dingo information sign. Although slightly overgrown in parts, an easy scenic walking track encircles the lake. (Note: the camping area is no longer open).

## What to look for

Getting there is half the fun as the drive takes in some beautiful scenery. Heading inland, it's possible to see the transition from an open vegetation type to a more complex, luxuriant rainforest type nearer the island's centre.

Lake Allom's plant life is interesting and varied and includes an original population of blackbutt trees. Look for the distinctively blackened (burnt) trunk as a way of recognising this eucalypt species.

On the walk down to the lake, visitors can see some quite large examples of kauri pines and satinays as well as cycads - an ancient group of plants dating back to the age of dinosaurs.

A variety of birds utilise the lake and its surrounding vegetation, including kingfishers. Large monitor lizards are often spotted around the carpark.

# Lake Allom

*Turtles are often seen breaking the water's surface.*

Freshwater turtles known as Kreft's river turtles inhabit the lake and are often seen close to shore, sheltering in nearby reeds and rushes. Naturally cautious, these creatures may at first be reluctant to appear, especially if people are swimming or making noise. But a patient observer is almost always rewarded.

As if on cue, several turtles will often emerge at once. Look for their dark outlines against the clearer patches of sand close to shore. A closer look may even reveal a turtle's prominent pale yellow streak along the side of its head or whether it is male (longer tail) or female (shorter tail).

Avoid feeding or handling the turtles. Although the acidic lakes they inhabit are relatively devoid of plant and animal life, these creatures are able to survive by feeding on insects that fall on to the water's surface. In times of hardship, they are capable of moving over land and relocating to other lakes.

## What we think

Lake Allom is a quiet, magical place, rarely visited by bus groups. The lake's turtles delight visitors of all ages and offer a chance for children to learn how to observe wildlife in a responsible way.

Swimmers should note that although the water is cool and clear, there is no beach for sunbathing. Also, because of its small size and vulnerability, make every effort not to pollute the lake or interfere with its delicate ecology. Use the toilets provided and don't allow soaps, detergents, suntan oils or insect repellents to enter the water.

Be prepared for a visit to Lake Allom (at a comfortable pace) to consume half a day. The tracks are quite rough and bumpy but, in our opinion, very scenic and well worth the effort.

# Central Fraser Island

**Map reference**: L7 / M7 (Hema), G6 (Sunmap)

**Location**: 12.5 km from Lake Allom, 23 km from Happy Valley

**What's there?** Walking track to lake, a sense of isolation

**Nearest to**: Boomerang Lakes, Lake Allom, Yidney Scrub

## Why go?

The Coomboo and Hidden Lakes Forest Walk offers some interesting scenery and a pleasant way to experience the island's diverse, natural interior.

With its tannin-stained water and diversity of plant life, Lake Coomboo (the larger of Coomboo Lakes) attracts walkers keen to escape the crowds and discover something new. Although not appealing as a swimming lake, Lake Coomboo is an interesting place to take photos and appreciate nature up close.

After a short back-track, the walk continues on to Hidden Lake, which, as its name implies, is another secluded freshwater oasis hidden in the forest. This longer walk is a popular choice with more serious hikers. Even so, the chances of encountering another soul are remote and may be part of the attraction.

Scientists have conducted some interesting research in the area. Based on radiocarbon dating, the organic deposits in Hidden Lake have been estimated to be at least 10 000 years old.

Scientific investigations have also confirmed the lake is perched above the main watertable (a 'perched' lake). The lake catches virtually no runoff and is topped up by rainfall instead.

## Getting there

It is only possible to access Lake Coomboo on foot via the Coomboo and Hidden Lakes Forest Walk. The shortest way in is from Northern Road, taking the signposted 400 m drive to the carpark. From here, it's a 2.2 km walk to Lake Coomboo. More detailed information on walking routes can be obtained from QPWS. The complete walk, starting from the carpark at the Lake Coomboo end and finishing at Northern Road at the other (the latter is signposted with a small parking bay), is 12.4 km one way and can take a few hours. Some walkers arrange to be dropped off at one end and collected at the other. Note, however, that the track to Lake Coomboo can be overgrown in parts and much of the vegetation is prickly. In fact, at the time of writing, the walking track to Hidden Lake was impenetrable.

## Facilities

No facilities are provided at any of these lakes, although a small carpark is provided at the start of the walk (Lake Coomboo end). Visitors should be self sufficient.

## What to look for

Lake Coomboo is surrounded by dense aquatic plant life, giving it an appearance somewhat reminiscent of Ocean Lake.

Old, twisted paperbarks occur along the edges. Underneath, the plant life is highly diverse and well-adapted to the harsh, low nutrient conditions. Look for curly wigs (a sedge with curly green stems), sundews (a tiny carnivorous plant, usually red and glistening), and geebungs (a larger shrub with subtle yellow flowers).

The dense vegetation supports a variety of ants and other insects. Birds are also abundant. Keep an eye out for bar-shouldered doves which sometimes fly across the tracks, while robins, wrens and honeyeaters flit through the understorey.

# Lake Coomboo

*Lake Coomboo's bleak, untamed beauty will appeal to some - not all - people.*

The lake itself is home to freshwater turtles, although these are generally shy and tricky to spot. Small native fish like the soft-spined sunfish and firetail gudgeon also occur in the area.

## What we think

Walking is an excellent, low impact way for some people to explore the island and experience its sights, sounds and smells.

Remember though, that walks like this are relatively isolated so some planning and reasonable fitness are required. While the terrain is not difficult, conditions can be hot and exposed at times and the sand a little soft in places. Avoid hot weather.

We've seen a few families walk to Lake Coomboo; the children tending to look unimpressed. The walks at Central Station and the short walk onto Wungul Sandblow may hold more interest for them.

Nature lovers like us will enjoy the lake.

113

# Central Fraser Island

**Map reference**: L6 / M6 (Hema), G5 (Sunmap)

**Location**: , 10 km from Yidney Scrub, 14.5 km from Lake Allom, 18 km from Moon Point, 19 km from Happy Valley

**What's there?** Freshwater boomerang-shaped lakes

**Nearest to**: Lake Allom, Lake Coomboo, Yidney Scrub

## Why go?

At 130 metres above sea level, the Boomerang Lakes are the highest perched dune lakes in the world.

In spite of this claim to fame, few visitors take the time to explore this unusual place. Its beauty and serenity often come as a pleasant surprise to those who decide to make the stop while exploring the island's scenic centre.

*In dry conditions, sections of track can become soft, requiring extra momentum*

With its white sandy beaches and diverse flora and fauna, the lake holds special appeal for naturalists, photographers and bushwalkers.

Only the southern lake is accessible from the carpark. Both lakes are long and skinny, roughly following the curve of a boomerang. Their unique shapes are best appreciated from the air.

The fact that the island has numerous perched lakes formed part of the island's case for World Heritage listing. Perched lakes are created when rainwater collects on top of an impermeable lens-shaped layer made of compressed sand and organic matter. The result is a lake filled entirely with rainwater, perched up above the main water table (in this case a long way up!)

## Getting there

The Boomerang Lakes are most commonly approached from the direction of Moon Point, Lake Allom or Yidney Scrub. The latter two form part of the Northern Forests Scenic Drive which, like other inland tracks, can become quite soft and slow-going in dry conditions. As you near the turnoff (heading northwards), you can catch some great glimpses of the lake from the road. From the Boomerang Lakes sign post, it's a short drive down to a very small carpark, followed by a short, easy walk down to the southern lake.

## Facilities

There are no facilities apart from the small carpark. The nearest settlement is located on the other side of the island, at Happy Valley.

## What to look for

The tea-coloured organic layer that has accumulated around lake's edge looks stunning against its white sandy beaches. It's possible to walk along the side of the

# Boomerang Lakes

*Shaped roughly like a boomerang, the lake's striking beauty and unspoilt qualities are sure to please visitors interested in nature.*

lake, following what appears to be an old vehicle track.

From the beach, just follow the lake around to the right. Nature enthusiasts will discover a wonderful diversity of plant life including sedges, grass trees, paperbarks and small carnivorous sundews.

The keen observer may spot a freshwater turtle in the water or come across a carapace and skeletal remains. Classified as a kind of short-necked turtle, the island's turtles might in fact represent a new species, quite different to Kreft's river turtles that occur on the mainland. While no fish have been recorded for the southern lake, scientists have confirmed the presence of firetail gudgeons in Boomerang North.

A little way along, you'll come to what at first may look like an island. This peninsula of vegetation juts out into the lake, helping to create its boomerang shape. From here, the lake may suddenly seem much larger - or at least longer - than it first did.

While there are no designated walking tracks, it is possible to walk along the beaches around most of the lake, so long as you take care to avoid damaging the fragile plant life (and watch for snakes).

## What we think

This quiet, secluded spot is strikingly beautiful and somewhat mysterious - it is easy to imagine the existence of undiscovered species lurking in its waters!

If you are in the vicinity and you're a nature lover, a stopover is a must. We're sure you'll agree that the lake's delicate ecology deserves protecting so take care to minimise impacts. Chances are, you will have this unusual place to yourself.

# Central Fraser Island

**Map reference**: L4 (Hema), G4 (Sunmap)

**Location**: 13 km (via beach) from Woralie Creek, 18 km from Boomerang Lakes, 32 km from Happy Valley, 49.5 km from Kingfisher Bay

**What's there?** Beach camping, stunning sheltered beach, great fishing

**Nearest to**: Boomerang Lakes, Bowarrady Creek, Woralie Creek

## Why go?

Sheltered from the dominant south-easterly winds, the beach just north of the Moon Point barge landing is one of the most stunning beaches on the island.

In calm conditions, the white sandy beach and idyllic blue-green water create the perfect spot for fishing, dolphin spotting or watching the sun set over the ocean.

*Sunset over Hervey Bay*

## Getting there

Vehicular barges depart from Urangan Boat Harbour (Hervey Bay) on the mainland and land at Moon Point and nearby Sandy Point (see p 17-18 for operating times and booking details). Concrete ramps are provided at both landings.

Being a barge landing area has the advantage of ensuring the main island tracks to Moon Point are well signposted. The drive, however, can be slow going, particularly if conditions are very dry and the sand is soft. Moon Point Road is generally in better condition than the side branching routes and continuing on towards Happy Valley, the drive through Yidney Scrub is very scenic.

Beach driving is not possible on the island's western side anywhere south of Moon Point. Heading northwards, vehicles are permitted on the beach as far as Wathumba Creek, but drivers should be extremely cautious, particularly at creek crossings and should not attempt to cross Wathumba Inlet.

If crossing Coongal Creek, apply the same caution as you would for other major creeks (see Eli Creek p 124-125 and Creek Crossings p 26-27). Coongal Creek releases quite large volumes of water and has steep sides in places. Creeks may be impassable, even at low tide. It is a good idea to get out and walk crossings first, checking for things like depth, buried weed and quick sand.

## Facilities

There are no facilities provided at Moon Point. The nearest, including a pay phone and general store, are located at Happy Valley on the opposite side of the island. Free-range beach camping is permitted behind the first dune, except where signposted otherwise. Boats can anchor near the mouth of Moon Creek at Sandy Point.

## What to look for

If you have a pair of binoculars, it's worth bringing them along. Pods of bottlenose dolphins are common in these waters and humpback whales are often seen in Platypus Bay from July to October as they journey to and from Antarctica. On a clear day, it's possible to spot their distinctive 'blows', even when kilometres out to sea.

Snorkellers occasionally take to the clear, calm waters which are home to a variety

# Moon Point

*With its white sand and blue-green water, it's hard to find a more beautiful setting for walking, fishing or camping than Moon Point.*

of fish (although there are no life guards on hand).

Nearby swampy areas can make biting insects a problem. The year's cooler months may be a better time to explore this part of the island when you are able to wear more clothing.

Campers can avoid the fragile dune vegetation by keeping to previously used tracks and campsites. Elsewhere, the sand is held in place by native groundcovers and leaf litter shed by coastal sheoaks. These 'pioneers' of the beaches and dunes are susceptible to damage.

## What we think

Moon Point's blue-green, sparkling waters and long stretch of white sandy beach are known to evoke sighs of longing from the passengers of passing whale watch boats. Surprisingly, most people coming to or from the island by barge pass over this beautiful spot in their hurry to get elsewhere.

We should warn you - biting sandflies and mosquitoes can quickly turn your stay in paradise into a bit of a nightmare. So come prepared, particularly if you are sensitive to insect bites.

*View from a whale watch boat*

117

# Central Fraser Island

**Map reference**: M8 (Hema), G6 (Sunmap)

**Location**: 4.5 km from The Pinnacles, 5.5 km from Lake Allom

**What's there?** Huge sandblow, lookout with spectacular views

**Nearest to**: Lake Allom, Maheno Wreck, The Pinnacles

## Why go?

Deriving its name from an Aboriginal knifeblade recovered on the surface of the dune, Knifeblade Sandblow is one of the largest, most impressive sandblows on Fraser Island. This desert-like island of sand stretches inland for more than three kilometres, dumping thousands of tonnes of sand onto the forests in its path.

*The track to Knifeblade*

From the lookout, it becomes obvious that the island is still evolving. The prevailing winds are dramatically shaping and reworking the landscape.

Sandblows are common on the island, beginning with a weak point in the dune closest to the beach. Wind funnels through forming a 'blow-out' that can gain momentum, becoming a large moving sand front or 'sandblow'.

On the outer margins of the dune, vegetation slows and traps the sand, creating what's known as its 'trailing arms'. The sand blow takes on a characteristic V or U shape which is best appreciated from above. When the distance becomes too much and the wind dissipates, the dune begins to slow. At last, the sea of inland vegetation is able to hold its ground and may gradually reclaim the dune. From the air, the impression of a U or V is still apparent in these ancient dunes.

Across the island, young through to ancient sandblows overlap, overlay and integrate with the greater sand mass, creating a complex picture of the island's development.

## Getting there

The turnoff for the lookout is located 4km inland from the island's eastern beach, along Woralie Road. The best time to drive on the beach is generally at low tide. Drivers can enjoy a scenic circuit drive by continuing along Woralie Road and connecting with Northern Road. There are some great locations along the way (eg Lake Allom, Boomerang Lakes, Yidney Scrub.) As with other inland tracks, prolonged dry spells or a lot of rain can cause conditions to deteriorate quickly.

Some great examples of scribbly gums, saw banksias and strangler figs can be seen along the track from Lake Allom. Keep an eye out for the small signs that will help you identify them.

From the signposted turnoff for the Knifeblade Sandblow lookout, it's just 400 metres to the carpark followed by a short walk to the viewing area.

## Facilities

A small but adequate carpark is provided as well as a timber viewing area with

# Knifeblade Sandblow

*Possibly the largest and most spectacular sandblow on the island, this great island of sand surges forth in a sea of vegetation.*

seating. There are no toilets or any other facilities.

## What to look for

Massive foxtail ferns flourish around the edge of the carpark and along the path to the lookout. There are plenty of midyim berries and grass trees as well as a beautiful big scribbly gum full of animal hollows. Another impressive tree is the smooth-barked apple gum, with its dimpled, red-stained trunk.

The view from the lookout takes in the spectacular interface between a tall, dense forest and the steep, unstable cliff face behind, across the top of the sand mass towards the distant ocean.

Leafless stands of sand-blasted trees await burial in front of the advancing sandblow, while on top of the dune, weathered branches and trunks return to the surface. These and any other remains, including Aboriginal implements, are later swept back into the great churning wall of sand.

## What we think

While all of Fraser Island's sandblows are impressive, Knifeblade really is hard to beat. This is island-shaping in action.

You can literally see the sheets of sand moving over the dune and spilling over the edges as the sandblow advances. It's a great reminder of the naturally dynamic nature of sandy coastlines. Those of us familiar with highly developed coastlines often forget that beach and dune stability has usually been artificially imposed on these areas (and takes a lot of upkeep).

Another bonus is that unlike many of the island's other star attractions, the lookout isn't crawling with tourists. In fact, most of the time there isn't a soul around.

# Central Fraser Island

**Map reference**: M8 (Hema), G6 (Sunmap)

**Location**: 2 km from Maheno Wreck, 5 km from Eli Creek

**What's there?** Coloured sand cliffs, beach fishing, ocean scenery

**Nearest to**: Cathedral Beach, Knifeblade Sand Blow, Maheno Wreck

## Why go?

Of all the coloured sand formations in the Cooloola sandmass, The Pinnacles are considered by many to be the most spectacular. In fact, some people claim they are the most impressive coloured sands in Australia.

Ocean winds and rainfall eat away at the formations, which are in fact little more than exposed section of dunes with layers of varying mineral content. Subtle yet powerful, rain splash and running water slice into the cliff face, carving out pinnacles, spires and other intricate shapes that appear to have been masterfully sculpted. Silts and clays hold the formations in place.

The Pinnacles are a site of cultural significance to the island's Aboriginal people, the Butchulla. In the Butchulla account of how the coloured sands were created, Wiberigan (the rainbow) is struck by a boomerang. The events leading up to this are beautifully described and illustrated on the visitors' sign. It's a fascinating story of love and revenge.

Today, visitors marvel at the colourful mix of reds, yellows, whites and variety of shades in between. Where iron oxides concentrate, the dunes appear very red. According to artists, it is possible to see more than seventy colours.

Elsewhere along the island's eastern beach, formations at Rainbow Gorge, Cathedral Beach and Red Canyon provide a similarly colourful glimpse of what lay beneath the island's sandy surface.

## Getting there

The Pinnacles are located on Seventy-Five Mile Beach, a short distance north of the Maheno Wreck. Visitors park high on the beach and follow the short walk into the chasm. Beach driving should only be attempted when the tide is suitably low and care should be taken when crossing Eli Creek (see Driving on Fraser Island p 22-35).

## Facilities

An information board is the only facility provided. (Rubbish bins are located just down the beach). The formations are fenced off to prevent visitors from taking samples and damaging the fragile dunes.

A fish cleaning facility (signposted) is

# The Pinnacles

*Arguably the most impressive coloured sands in Australia, The Pinnacles reflect the unrelenting artistry of ocean, wind and rain.*

provided on the beach just north of The Pinnacles with important information on how and where to clean your fish and how to dispose of the waste.

The K'Gari Camping Area is located just inland, a few kilometres south of The Pinnacles (see p 11).

## What to look for

Early morning, as near as possible to sunrise, is the best time for viewing and photographing the coloured sands. Clear blue skies can create a striking contrast with the yellows and golds of the dunes. The colours can also appear more vivid after rain.

You'll have to resist any temptation to take samples of the coloured sands as even slight damage can lead to landslides. For the same reason, keep well away from the cliffs and dunes. Do not allow children to play near any of the island's sand cliffs, as landslides can happen unexpectedly and silently.

Dingoes are sometimes seen patrolling the beach in search of dead birds, fish and food scraps. It is important not to feed or interact with the dingoes. Offshore, humpback whales and dolphins may be spotted and up above, kites and other birds of prey may be seen. Beach fishing is popular along much of this beach.

## What we think

The Pinnacles are one of Fraser Island's natural treasures and an essential stop along the beach. This is the kind of place most photographs can't do justice to - the beautiful shapes and colours have to be seen to be believed. If you are determined to take a good photo, be prepared for an early start.

# Central Fraser Island

**Map reference**: M8 (Hema), H6 (Sunmap)

**Location**: 3 km from Eli Creek, 9 km from Happy Valley

**What's there?** Shipwreck on beach, beach fishing, wild ocean scenery

**Nearest to**: Eli Creek, Knifeblade Sandblow, The Pinnacles

## Why go?

Rising impressively out of the mist, as if to defy the constant battering of surf and saltspray, the famous and popular Maheno stands as a rusting reminder of the area's history of unpredictable seas and weather extremes.

The wreck provides opportunities for photography, nearby beach fishing and taking in the sights and sounds of the surf beach environment. It's also one of the few man-made historical attractions on the island.

Believe it or not, this rusty relic was once a luxury steam ship, capable of transporting nearly 300 passengers, including over 200 First Class. Built in Scotland in 1905 for a New Zealand steam ship company, the Maheno mostly ferried passengers across the Tasman Sea. Fitted with chandeliers and a grand piano, she was something like the Titanic in her day, except on a much smaller scale. Weighing in at 5323 tons and capable of speeds of up to 25 knots, the Maheno was very quick for her time and made record runs between Auckland and Sydney. During the First World War, she served as a hospital ship and when it came time for retirement, was sold to Japan for scrap metal.

In 1935, the Maheno was being towed up the coast from Sydney by a veteran steamer called the Oonah when an unseasonal cyclone hit. The tow rope snapped and the Maheno helplessly drifted onto Fraser Island's eastern shore, eventually settling just north of Eli Creek. After attempts to refloat her failed, the ship was abandoned and her furnishings auctioned off on the beach to recoup some of the monetary loss. Her wartime service was revived when, in World War II, she was used as a target for bombing practice by the Royal Australian Air Force.

Today, the Maheno looks remarkably well considering the battering from the elements and destruction of the bombing raids. Sand now covers much of the wreck, leaving only the uppermost part of her three decks exposed. Some of the remains of the timber decking are still in tact. Each year, more and more of the wreck disappears.

## Getting there

The Maheno is situated on Seventy-Five Mile Beach, just north of Eli Creek. When driving on the beach, drive as near as

# Maheno Wreck

*One of Fraser's most well-known landmarks, the Maheno takes a battering from the waves, sand and salt-laden winds.*

possible to low tide, and apply caution when crossing Eli Creek (see Driving on Fraser Island p 22-35). For the safety of visitors, be sure to slow down and exercise caution when approaching the wreck. Once out of your vehicle, remember that the wreck is essentially situated on a highway and the sound of the surf often muffles the noise of approaching traffic.

## Facilities

Interpretive signage, rubbish bins and warnings about the dangers of climbing on the wreck are the only facilities provided. Eli Creek has the nearest toilets.

## What to look for

From a distance, the sight of the Maheno rising out of the sand and nearby surf can be quite spectacular. On a clear day, you may be able to spot it from several kilometres off as you drive up the beach. But the wreck is impressive in any weather. Storm clouds, mist and seaspray create a wild and almost ominous sight sure to please photographers and romantics alike.

Keep an eye out for dingoes, whales (between July and October) and birds of prey. Sharks may also be seen near to shore (it is not considered safe to swim).

## What we think

If you're planning to explore the island's eastern coastline, this beautiful shipwreck is a 'must see'. Be sure to stop and take some photos. Late afternoon - when the sun's out - is often best (tides permitting).

You can gain an appreciation of the Maheno's former self by taking a look at some of the old photographs on display at Happy Valley and the ship's anchor outside Kingfisher Bay Resort. A small-scale replica is also on show at Kingfisher's Maheno restaurant.

# Central Fraser Island

**Map reference**: N8 (Hema), H6 (Sunmap)

**Location**: 6 km from Happy Valley, 26 km from Eurong, 34 km from Kingfisher Bay

**What's there?** Clear freshwater creek, board walk, ocean beach

**Nearest to**: Happy Valley, Maheno Wreck, The Pinnacles

## Why go?

Eli Creek is the largest creek on the eastern side of the island. It offers great opportunities for shallow swimming, wading, walking and nature appreciation.

The creek springs from an underground aquifer more than five kilometres inland, pouring around 80 million litres of fresh water into the ocean each day. Naturally chilled and filtered, the emergent water passes silently over a bed of clean, quartz sand before finally being lost to the pounding surf and seawater.

Exploring the creek from the water is a great way to cool down and in most places the water is just a few feet deep. Visitors simply hop in upstream and let the creek's current gently carry them along for a hundred metres or so, meandering around natural bends and lushly vegetated banks.

## Getting there

Eli Creek is situated on the island's eastern beach. Parked cars and buses usually make the creek easy to find.

When crossing Eli Creek, always try to cross at its shallowest point (nearest the sea), preferably at low tide. Use a low gear and a little momentum (see Creek crossings p 26-27). Remember that the creek can regularly alter its size and course and may have changed since you last saw it. Rapid changes are also caused by ocean-moved sand near the mouth of the creek. It is not advisable to attempt a creek crossing at night.

## Facilities

Only the last section of Eli Creek is available for public use. Non-swimmers can enjoy a short, interesting circuit around part of the creek aided by interpretive signage and wooden boardwalks. Toilets (with wheelchair access) are situated within walking distance.

Drinking water is best taken upstream from where visitors swim and wade (although boiling or other treatment is recommended by park authorities).

## What to look for

All sorts of creatures inhabit Eli Creek. Eels, tiger perch and whiting are often spotted and are quick to take cover in the luxuriant vegetation. A variety of native shrubs, sedges and rushes help to protect the creek's banks from erosion and provide shade and habitat for native animals.

*Take care crossing the creek mouth*

Look for the distinctive orange fruits and above-ground 'stilt' roots of the pandanas palm. Local Aborigines used sophisticated methods to extract and prepare the edible inner seeds. When eaten raw, the fruits gave experimental explorers chronic diar-

# Eli Creek

*Popular with swimmers, Eli Creek delivers a steady stream of cool fresh water amidst a jungle of interesting plant life.*

rhoea, sore lips and blistered tongues.

High above, birds of prey circle the skies while the beach is home to patrolling dingoes and interesting birds like the pied oystercatcher and crested tern.

If planning to swim, remember the creek's current is quite strong and there are no life guards. Watch for exposed tree roots and stumps. Other possible hazards include dingoes and traffic on the beach (including planes) and in the soft sand. It's also a good idea to stay clear of cliff faces in case of sand collapse.

## What we think

If you're prepared to get wet and share the creek with its steady stream of visitors, you'll immediately understand why this beach oasis is an essential stop for many Fraser Island regulars. In hot or cold weather, we find it a great place for a refreshing dip. The constant squeals of delight suggest it's a holiday highlight for water-loving kids, some coming prepared with tyre tubes and boogie boards. Where the creek widens out, younger children and their parents enjoy the shallower, slower-running water. (Note: sharks, rips and dangerous surf make nearby ocean waters unsafe for swimming).

This will seem obvious - but make sure you actually venture up the creek. We've seen a few people splash themselves with water where the cars are parked, thinking they've 'done Eli', then leave!

# Central Fraser Island

**Map reference**: N7 (Hema), H5 (Sunmap)

**Location**: 13 km from Happy Valley, 22 km from Moon Point

**What's there?** Stunning rainforest, scenic driving

**Nearest to**: Boomerang Lakes, Happy Valley, Lake Garawongera

## Why go?

Arguably as beautiful as Central Station, Yidney Scrub is an impressive pocket of rainforest located in the island's centre. Home to a virgin stand of Kauri pines and variety of other rainforest species, this destination appeals to botanists, photographers and nature lovers alike.

Most visitors, however, first discover Yidney Scrub on their way to or from Moon Point, unprepared for the vegetation's dramatic change from open woodland and flowering heath, to a luxuriant, complex forest type.

*Carrol (Backhousia myrtifolia)*

## Getting there

Yidney Scrub often comes as a pleasant surprise to those crossing the island. Visitors keen to explore the island's centre can continue along a scenic circuit drive that takes in Boomerang Lakes, Lake Allom and Knifeblade Sandblow. This 38 km drive can be started from Happy Valley or just south of The Pinnacles and takes 2.5 hours or longer, depending on the amount of time spent at places along the way and condition of the tracks. There is usually limited traffic. In very dry or very wet weather, the tracks can be slow-going - for instance, the sand may be very soft or sections of track may be washed out.

As you pass through Yidney Scrub, small signs indicate where the rainforest begins and ends. Continuing on towards the west coast, the forest trees are replaced by cycads and blackbutts until open woodlands and spectacular healthlands with banksias and foxtails emerge.

## Facilities

There are no facilities located at Yidney Scrub, apart from small signs indicating the species names of some of the trees and a few large passing bays that can be used for a quick stop.

## What to look for

As you drive into Yidney Scrub, the vegetation suddenly changes. The canopy closes overhead and the temperature drops noticeably, plunging you into a cool, dark rainforest.

Satinays, brushboxes, carrol, piccabeen palms and kauri pines are abundant but keep an eye out also for strangler figs - there are some impressive specimens that can be spotted from the road.

Kauri pines are easily recognised by their tall, very straight trunks. One very large example can be seen on the side of the

# Yidney Scrub

*Superb rainforest, shade and a cooler microclimate make the scenic drive through Yidney Scrub a tempting option on a hot day.*

road. The first species logged on the island, their trunks are said to have made excellent ship masts. There are also plenty of very big brushboxes, distinguishable by their scaly bark, as well as the island's signature species - the Fraser Island satinay (see Pile Valley p152-153).

Mosses and lichens grow in thick carpets over many of the trunks, especially those of trees that have fallen. Termites and microorganisms slowly break down these logs and in doing so, liberate nutrients for reuse by the rainforest.

## What we think

Although part of a drive, this stunning patch of rainforest deserves to be thought of as a destination in its own right. A drive to Yidney Scrub is perfect for people who don't have time to visit Central Station (or wish to steer clear of the hordes of cars and buses) but still want a look at some of the most beautiful rainforest that Fraser has to offer.

It's a great place to escape to on a hot day and well worth a look, regardless of whether you choose to continue along the tourist drive. We often stop here for a quick cup of tea.

*Strangler figs grow from the top down, using their host tree for support*

# Central Fraser Island

**Map reference**: O7 (Hema), H5 (Sunmap)

**Location**: 8 km from Happy Valley, 30 km from Moon Point

**What's there?** Freshwater lake, white sandy beach, lots of birds, picnic areas

**Nearest to**: Happy Valley, Rainbow Gorge, Yidney Scrub

## Why go?

Clean, fresh water and white sandy beaches help to explain Lake Garawongera's popularity as a swimming and picnicking destination. Nature lovers are also well catered for, with the forests encircling the lake teeming with birdlife.

Just like the more famous Lake McKenzie, this is a unique type of lake known as a 'perched' lake. It is essentially a large body of rainwater, perched up above the island's main water table. A hard layer of compressed sand and organic matter has formed a lens underneath the lake, without which, the cool, clear water would seep through the sand.

## Getting there

From Happy Valley, the scenic drive towards Lake Garawongera passes through some beautiful rainforest. Although very pleasant, note that the 8km trip is very bumpy, slow-going and can be soft. Good vehicle clearance may be required. From the west, the lake can be reached via Happy Valley Road which can also be soft and slow-going. The carpark turnoff is signposted.

## Facilities

A good-sized car park, toilets (wheelchair access), interpretive signage, barbecue and sheltered picnic facilities are provided. Happy Valley is the nearest township and has a pay phone, general store and more comprehensive facilities.

In order to reach the lake from the car park, visitors must follow the short walking track down to the lake shore.

## What to look for

In the shallows, firetail gudgeons (a small freshwater fish) can often be spotted. But although the lake may appear to offer ideal habitat for high numbers of species, surprisingly few animals can survive here.

Much like a treated swimming pool, the naturally acidic conditions discourage the growth of plants and algae which, if able to flourish, would provide food and habitat for a greater abundance of fish and other wildlife.

Species able to handle the conditions are often described by scientists as 'most unusual'. In lakes like this - on the sand islands off southeast Queensland - such creatures include freshwater turtles, dragonflies, damselflies and a few fish.

Other oddities include an ancient freshwater sponge and unusual types of water fleas, mites, flies, worms, spiders and midges. Several new species have been discovered, including a magnificent dark blue dragonfly.

*Dragonflies are never far from water*

# Lake Garawongera

*Lake Garawongera offers a quiet escape for nature enthusiasts and water lovers.*

Dragonflies are well worth looking for - the stands of native aquatic plants in and around the lake provide an ideal habitat for them. Some scientists have suggested that dragonflies and damselflies could be used as bioindicators for the 'health' of dune wetland areas such as this.

Using two sets of wings and a clever mouth design, these fascinating insects are able to hunt and devour their prey mid flight!

Musk ducks are sometimes spotted. Look for their strange, partly submerged shapes in the water. Males have a large, black lobe of skin hanging under their bill and court the females using spectacular splashing displays. Both are expert divers but are rarely seen taking to the air. While you might spot one or two of these birds, don't expect to see large flocks of ducks. Like other perched lakes on the island, the productivity of the lake is too low to support large numbers.

In the surrounding forests, expect to see goannas, kookaburras, and a myriad of smaller birds.

## What we think

With its crystal clear water and peaceful, natural surroundings, Lake Garawongera is right up there with Fraser Island's most beautiful lakes.

To avoid upsetting the lake's delicate, natural balance, keep fruit skins, shampoos, insect repellents and other pollutants from entering the water. Use the toilets provided and take care not to leave cigarette butts behind. Sadly, cigarette butts collect around the edges of some of the more popular lakes, polluting and persisting in the environment.

# Central Fraser Island

**Map reference**: O7 / O8 (Hema), J6 (Sunmap)

**Location**: 8 km from Lake Garawongera, 20 km from Eurong

**What's there?** Seaside township & resort, beach fishing, good facilities

**Nearest to**: Lake Garawongera, Rainbow Gorge, Yidney Scrub

## Why go?

Happy Valley's relaxed atmosphere and good facilities attract a range of visitors, from adventure-seeking backpackers to families and retirees.

Since its 1930s beginnings, this small resort village has been a popular base for beach fishing. Today, Happy Valley can be quiet one minute and packed with tour groups the next. As with the other island townships, excited backpackers often transform this spot into a sociable, cosmopolitan hub of activity.

There is plenty to see and do in the vicinity. Within an easy distance are the colourful sands of Rainbow Gorge, beautiful inland forests of Yidney Scrub and the famous Maheno Wreck. Beach lovers often take to the cold, flowing waters of Eli Creek or journey inland to relax on the shore of Lake Garawongera.

For those who choose to stay over, Happy Valley allows visitors to experience the wilderness qualities of the island's eastern beach in relative comfort.

## Getting there

Happy Valley can be accessed from a number of directions. Driving up or down Seventy-Five Mile Beach is the smoothest and often most direct route. Check tide times before leaving and, if approaching from the north, take care crossing Eli Creek (see p124-125, 26-27). Planes take off and land on the beach between Happy Valley and Dundubara, just over 20 km to the north so be alert and keep well clear.

From the barge landing at Moon Point, the drive across the island to Happy Valley is well signposted but in dry conditions, can be soft and slow-going. The last section, through Yidney Scrub, is shaded and very scenic. Another scenic inland route is via Lake Garawongera. Again, allow plenty of time as the track can be soft, bumpy and narrow in parts.

Happy Valley also serves as a starting (or finishing) point for bushwalkers undertaking the Fraser Island Great Walk or sections of it (see p14-16).

## Facilities

Facilities include a public pay phone, ambulance service (operates in peak periods), toilets, drinking water, general store (petrol, diesel, ice, liquor, eftpos), bistro and bar, reception and tour bookings, guest swimming pool, BBQ area and vehicle recovery and towing service.

Known as Fraser Island Retreat, the resort located at Happy Valley provides nine self-contained lodges. Advance bookings are recommended as tours and accommodation can fill up, even outside of holiday periods. In addition, there are numerous privately-owned holiday homes available for rent.

# Happy Valley

*With its relaxed holiday atmosphere, the township of Happy Valley - centrally located on the island's eastern beach - makes an ideal fishing and touring base.*

## What to look for

Just a short walk from the township, the beach provides great opportunities for fishing, walking and relaxation. Nature lovers enjoy spotting dolphins and it's not unusual to catch sight of a shark close to shore. Sharks and strong currents make the ocean unsafe for swimming. Yidney Rocks and the beach adjacent Happy Valley are popular fishing spots, especially during the tailor season (Aug-Sept).

Dingoes may be seen patrolling the beach or township. Don't be tempted to feed or encourage them (see Dingo safety p 45). Being natural scavengers as well as hunters, dingoes may be spotted devouring dead birds and fish they find washed up along the beach.

Humpback whales can be spotted, often some distance from shore, as they pass the eastern side of Fraser Island on their annual migration between Antarctica and north Queensland (July-October). Look for their distinctive 'blow' which can be seen for kilometres on a calm day.

Birdwatchers are also catered for. Birds of prey such as the whistling and brahminy kite are often sighted. Sea birds like the tern and pied oystercatcher can be spotted on the beach taking their chances with the traffic, while a variety of smaller bird species inhabit the native vegetation surrounding the township.

## What we think

Happy Valley is a charismatic seaside location well worth a look. For those who don't want to rough it, it's an excellent base from which to explore a range of Fraser Island attractions. The resort and café have a great holiday atmosphere.

# Central Fraser Island

**Map reference**: P7 (Hema), J6 (Sunmap)

**Location**: 4 km from Happy Valley, 17.5 km from Eurong

**What's there?** Multi-coloured gorge, sandblows, circuit walk, ocean beach

**Nearest to**: Eli Creek, Happy Valley, Lake Garawongera

## Why go?

In spite of its appealing name, Rainbow Gorge only manages to lure a handful of visitors off the island's popular eastern beach each day.

As well as the small colourful gorge, many visitors come for the circuit walk (approximately 2 km) around Kirrar Sand Blow. From the top of the sand blow, walkers are treated to panoramic views that sweep towards the ocean and outer edges of this impressive moving dune.

## Getting there

Situated a few kilometres south of Happy Valley, Rainbow Gorge can only be accessed from Seventy-Five Mile Beach. Beach driving should only be attempted when the tide is sufficiently low and conditions allow. Parking is on the beach, near the start of the walking track. A small sign indicates the start of the walk in.

## Facilities

A picnic area is all that is provided. The nearest amenities, including toilets and a general store, are located at Happy Valley.

## What to look for

The first part of the walk to the gorge passes through a cypress pine forest. Keep an eye out for orb weavers. These elegant-looking spiders string large webs between the trees.

The gorge gives the impression of having been painted from a palette of reds, browns and yellows with masterful splashes of white and black.

Snapping a great photo, however, can be a challenge since the gorge is often cast in shadow. The colours are usually at their best in the early part of the morning, especially on clear, sunny days.

Please resist the temptation to remove sand for a souvenir, no matter how small. This practice is now prohibited because of the damage it can cause to the dunes.

For those who continue along the circuit walk, there are some great views from the top of Kirrar Sandblow. Sandblows like this are a natural feature of the island's east coast, playing a vital role in shaping the sand mass. A small gap in the seaward vegetation can be enough to

*Only the toughest plants survive*

# Rainbow Gorge

*Ironstone outcrops display unusual patterns atop the sandblow.*

trigger a sandblow. Once a wind tunnel takes effect, increasing amounts of sand are funneled through, eventually creating a vast moving front.

At the top of the gorge, check out the unusual patches of ironstone - a type of red-coloured sandstone - on the surface of the sandblow. This is one of the few places on the island where ironstone can be easily viewed.

## What we think

Rainbow Gorge is a good option if you only have a day or two on the island but don't want to miss out on two of its stunning natural formations - coloured sands and sandblows. Here, both can be seen in the one place plus you have the perfect excuse to stop and stretch your legs before continuing your drive along the beach. Note that the walk to the top will seem like hard work for some people, especially if it's hot or windy.

*A colourful glimpse of what lies beneath the island's sandy surface*

133

# Central Fraser Island

**Map reference**: P6 (Hema), J5 (Sunmap - track not depicted)

**Location**: 17 km from Stonetool Sandblow, 18.5 km from Kingfisher Bay

**What's there?** Giant trees, rainforest

**Nearest to**: Kingfisher Bay, Lake Wabby, Stonetool Sandblow

## Why go?

Some of the island's biggest trees can be seen growing in the Valley of the Giants. At the height of the logging controversy, this was the site of anti-logging protests.

Today, this pretty pocket of rainforest is slowly earning an occasional spot on the itineraries of tour operators and past visitors to the island who are looking to explore somewhere new. The long drive appeals to those interested in massive trees, lush vegetation and serenity.

There is also a small but steady stream of botanists, conservationists, photographers and journalists who come to study or pay homage to the giant trees.

## Getting there

The beginning of the circuit drive can be reached from several directions. All can be slow-going (particularly in very dry conditions) and carry a limited amount of traffic compared with those linking the more popular destinations further south. These tracks are very scenic with some striking white stretches of sand and ever-changing vegetation.

A 2 km track leads to the 10 km circuit - both are classified as minor tracks and are not regularly maintained. Fallen trees can sometimes block the track and prevent you from completing the circuit, so check conditions before you leave. Rangers and local tour operators may be able to give you an update.

Bushwalkers can also access this location as part of the Fraser Island Great Walk from Lake Wabby (16.2 km one way) or Lake Garawongera (13.1 km one way). See pages 14 -16 for more information on walking options.

## Facilities

There are no major facilities provided in this area or in the immediate vicinity. The walkers' camp offers a toilet, platform seating and small tent sites (permit and booking required, see p 16).

## What to look for

The circuit drive will take you past some very large trees, the diameter of a few of these spanning several metres.

Amongst the most impressive species in terms of trunk diameter are the tallowwood (*Eucalyptus microcorys*), satinay

*The long circuit drive*

# Valley of the Giants

*Giant stumps like this are a reminder of the massive size of trees in days gone by.*

(*Syncarpia hillii*) and brushbox (*Lophostemon confertus*); all members of the Myrtaceae family.

The Giant Tallowwood is situated approximately 3.7 km from the start of the circuit, heading clockwise (or 2.6 km return from the walkers' camp). Further along, those keen for a walk can check out the Giant Satinay (approximately 5.3 km return). Bear in mind that leeches can be abundant here so keep to the track.

Evidence of the island's logging history occurs a short distance from the Valley of the Giants and includes the ruins of Poyungun Forestry Camp, Petries Camp and a log landing on the west coast.

## What we think

This location has trouble competing with the visual splendours of better-known places like Pile Valley and Yidney Scrub.

But if trees excite you and you have half a day or more to spare, a peaceful drive through the Valley of the Giants may be a worthwhile option for you.

However, in spite of what its name implies, this is *not* one of the island's best kept secrets. The circuit covers quite a large area, with only a few large trees visible from the road. Some people even leave wondering where the 'valley' was.

Still keen to go? Chances are you love trees (or bushwalking) and won't be disappointed. There are some stunning examples so don't forget to take your camera and perhaps even give one a big hug while you're there!

135

# Central Fraser Island

**Map reference**: P4 (Hema), J4 (Sunmap)

**Location**: 14 km from Central Station, 15 km from Lake McKenzie, 22 km from Eurong

**What's there?** Resort & village, scenic walks, nature-based activities, great facilities, diverse plant & birdlife

**Nearest to**: Lake McKenzie, Central Station, Basin Lake

## Why go?

Kingfisher Bay attracts those in search of a wilderness holiday with the added comforts of a luxurious island resort that blends into its natural surroundings.

Shopping facilities, walking tracks, guided tours and nature-based activities offered by the resort's own ranger staff attract large numbers of people from all walks of life, especially during the whale-watching season (July - October).

*A lookout offers magnificent views across the Great Sandy Straits*

## Getting there

From the mainland, separate barge and passenger ferries depart River Heads and Hervey Bay respectively, arriving directly at the resort village.

Vehicle tracks leading to Kingfisher Bay from major inland locations like Lake McKenzie, Central Station and Lake Wabby are signposted. Expect to meet other vehicles en route including large tour buses which will expect right of way.

While these relatively busy tracks are some of the most regularly-maintained on the island, conditions can deteriorate rapidly during very dry or rainy periods. Engage four-wheel drive as soon as you leave the bitumen, no matter what the conditions.

Hiking to Kingfisher Bay requires time, reasonable fitness and preparation. One of the more popular routes is from Lake McKenzie (23.5 km return). This forms part of the Fraser Island Great Walk (see pg 14-16). Be sure to obtain information and advice prior to leaving.

## Facilities

Provisions for day visitors and resort guests include a general store (with fuel, souvenirs, ice, eftpos etc), café / bakery, pay phones, toilets, resort rangers and tour booking office, playground, car hire, jetty kiosk (with bait, fishing equipment, boat hire), walking tracks and bistro / bar.

Other resort facilities include 24-hour reception, tennis courts, child care, beauty salon, swimming pools, spa, bars, conference areas and restaurants.

Accommodation includes hotel rooms, self-catering villas and share lodges. At times, the resort is fully booked so it is best to make a reservation in advance. Camping facilities are not available.

# Kingfisher Bay

*Kingfisher Bay Resort and Village offers a suite of nature-based activities, the calm waters of Hervey Bay adding to the area's serenity.*

## What to look for

Early morning bird watching, slide shows, dingo talks, bushtucker and flora walks, children's activities and nocturnal animal spotlighting are some of the activities conducted by resort ranger staff. They offer free fact sheets as well as information on walks to nearby places of interest such as Dundonga Creek and McKenzie's Jetty (the latter being a former World War Two Z force commando training camp).

Wildlife sighted by night includes dingoes, frogs, sugar gliders, bats, tawny frogmouths, carpet pythons and marine animals such as dolphins and shovel-nosed rays. Patience and good timing is required. A lucky observer may catch sight of energetic feathertail gliders, melomys or ground-dwelling bandicoots.

Birds are the most conspicuous animals by day. The diverse plant communities, which include coastal heath and wetlands, support an amazing variety of birds. Look for whistling kites, brahminy kites, welcome swallows, bar-shouldered doves, grey shrike-thrushes, rainbow lorikeets and various honey eaters, kingfishers and migratory waders. During spring, the heath is filled with colour, the multitude of flowers attracting insects, birds, blossom bats and other pollinators. Near the jetty, children love the great armies of soldier crabs that take to the beach at low tide.

## What we think

With so much to see and do, this beautifully-situated resort and village is a popular choice with families, couples, backpackers and others. But despite being a busy tourist hub, the resort's clever design makes it possible to feel as secluded or social as you choose.

This is one of Australia's top ecotourism resorts and well worth a look for this reason alone.

# Southern Fraser Island

Fraser Island's south is synonymous with beautiful lakes and stunning rainforest. This is where many of the island's famous beauty spots are concentrated (and where you'll also find the highest concentrations of tourists).

Here, you can explore seven of the freshwater lakes described in this guide book, including the famous Lake McKenzie.

With crystal clear waters and white sandy beaches, Lake McKenzie and Lake Birrabeen are popular with swimmers. Nearer the east coast, a giant sand dune slowly engulfs Lake Wabby, while further south, tea-coloured waters gently lap against the shores of Lake Boomanjin - the world's largest perched lake.

*Snout Point*

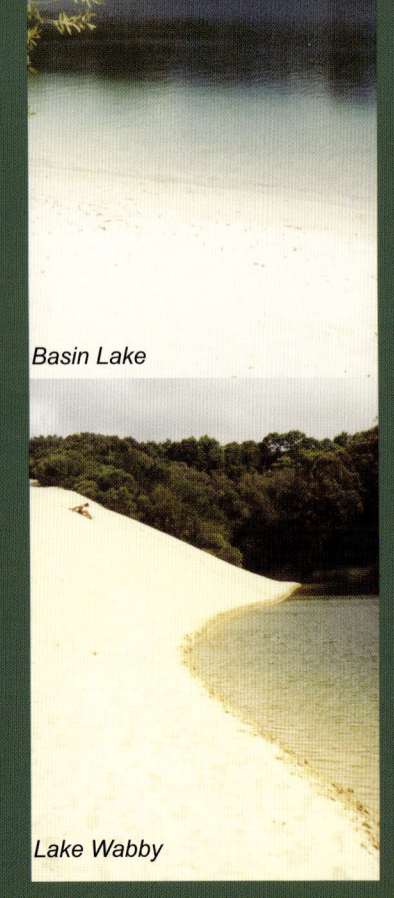

*Basin Lake*

*Lake Wabby*

*Central Station*

KINGFISHER BAY
angaroo
Island
River Heads
STRAIT
LAKE
McKENZIE
CENTRAL
STATION
STONETOOL
SANDBLOW
LAKE
WABBY
BASIN LAKE
PILE
VALLEY
Turkey
Island
UNGOWA
LAKE
BIRRABEEN
LAKE
JENNINGS
EURONG
LAKE
BENAROON
LAKE BOOMANJIN
DILLI VILLAGE
ISLAND
GARRY'S
ANCHORAGE
GREAT
SNOUT
POINT
SANDY
SOUTHERN
TRACK
N
W
E
S
STRAIT
HOOK
POINT
Tin
Can
Inlet

# Southern Fraser Island

**Map reference**: Q6 (Hema), K5 (Sunmap)

**Location**: 5 km from Lake Wabby lookout carpark, 9 km from Eurong

**What's there?** Impressive sandblow, lookout with ocean views

**Nearest to**: Eurong, Lake Wabby lookout, Rainbow Gorge

## Why go?

A source of fascination for nature lovers, photographers and visitors interested in cultural history, Stonetool Sandblow offers a timeless insight into the natural processes that have shaped the island for many thousands of years.

As sand sweeps over the surface of the dune on its relentless passage inland, Aboriginal implements are sometimes uncovered. Once used by the island's traditional custodians, the Butchulla, these items are a reminder of a healthy people who enjoyed an enduring association with the island, long before white exploitation.

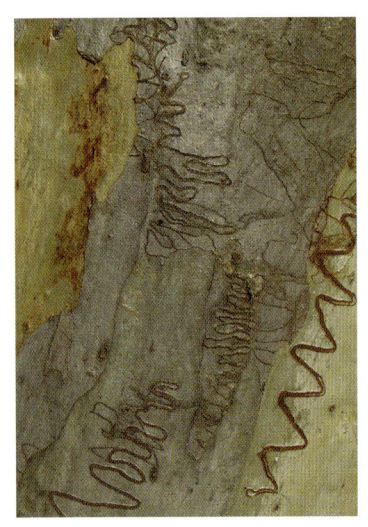

*Scribbly gums are easy to recognise*

Shaped by the prevailing south-easterly winds, Stonetool Sandblow is typical of the dozens of sandblows dotted along the island's east coast. As the sun shines directly overhead and then sets in the west, the dune is often illuminated then bathed in a beautiful golden light.

## Getting there

One of the attractions of the Central Lakes (Lake Wabby) Scenic Drive, the lookout is conveniently located a short distance from the eastern beach, on the way to Lake Wabby. From the beach, turn off onto Cornwells Break Road. A couple of kilometres along, take the Stonetool Sandblow turnoff on the right. This immediately leads you into a short one-way section and the carpark. The carpark has limited spaces but is usually very quiet.

Passing through sheoak and paperbark forests, the drive from the beach along Cornwells Break Road involves a very steep climb. In dry conditions, the sand can become extremely soft. It is not uncommon for vehicles, including buses, to become bogged or require several attempts at the ascent, which can worsen track conditions. Park authorities have installed sections of wooden boards to make the going easier. If traveling toward the beach, drive slowly down the hill as vehicles may be on their way up and there is limited room for passing.

## Facilities

The lookout is a short, easy walk from the carpark. Picnic tables are available for all visitors to enjoy, but are often used by bus groups in the morning and late afternoon. Courtesy is the key. Toilets are not provided here.

## What to look for

With spectacular views across this natural, mobile dune towards the South Pacific

# Stonetool Sandblow

*Shaped by the wind, Stonetool Sandblow is a natural 'island of sand' that buries and exhumes ancient forests and hidden treasures.*

Ocean, the lookout offers the perfect vantage point for taking photos and spotting wildlife. In particular, keep an eye out for birds of prey soaring overhead. Whistling and brahminy kites are often seen - to tell them apart, listen for the long descending call of the whistling kite followed by an upward 'si-si-si-si'. Don't' forget to bring your binoculars. There is a great view towards the ocean and, from around July until October, it may be possible to spot humpback whales in the distance.

Always moving and changing, the surface of the sandblow can be a fascinating sight. Dead trees that were once buried and killed by the advancing wall of sand re-emerge, their weathered branches taking on bizarre and intricate forms. Along the trailing arms of the sandblow, the protruding canopies of buried trees signal their desperate struggle to survive and possibly reclaim the dune.

Scribbly gums are plentiful and some great examples can be seen growing along the path to the lookout. The interesting bark patterns are created when the larvae of inscripta moths burrow and feed beneath the bark's surface, leaving scribbles as they go.

## What we think

A great example of a sandblow (and all visitors to Fraser should see at least one sandblow), Stonetool is definitely worth a look if you're in the vicinity. It's a quick and easy stop for those with limited time, thanks to its handy position and accessible lookout. Although a few bus groups visit the lookout, it's usually possible to get it all to yourself.

Our favourite time to visit is mid to late afternoon. The sun is behind you, making it ideal for taking photos and soaking up the phenomenal view.

# Southern Fraser Island

**Map reference**: R6 (Hema), K5 (Sunmap)

**Location**: 6.5 km from Eurong, 16.5 km from Lake McKenzie

**What's there?** Deepest freshwater lake, sandblow, walks, lookout with ocean views, freshwater fish

**Nearest to**: Eurong, Rainbow Gorge, Stonetool Sandblow

## Why go?

Lake Wabby is a picturesque, living example of the dynamic link between sand, sea, vegetation and freshwater. It offers unique opportunities for taking photos, walking, birdwatching and appreciating nature, and is a favourite spot for swimmers and the occasional snorkeller.

The advancing edge of Hammerstone Sandblow forms a steep bank or 'barrage' that is steadily moving inland, slowly filling the lake with sand. The colour of the sand contrasts starkly against the lake's deep green waters. With depth estimates usually in excess of 10 m, Lake Wabby is thought to be the island's deepest lake. These waters are home to at least 11 different species of freshwater fish.

## Getting there

Lake Wabby can be accessed by vehicle from Kingfisher Bay on the island's western side, following signs. This track has some steep downhill sections and can be quite soft. From the island's eastern beach, access is via Cornwell's Break Road. Turn left at the first intersection after Stonetool Sandblow and follow the signs to Lake Wabby Lookout. This section is quite narrow and overgrown, with limited passing bays. The lookout is just a short distance from the carpark. (The Fraser Island Great Walk connects with this - see p14 -16). From the lookout, a clearly marked 3.1 km (return) walking track leads down to the lake, requiring a reasonable level of fitness for the return trip. It is also possible to park on the eastern beach, adjacent to the lake, and take either the Ridge Track (4.7 km return) or Lake Wabby Circuit (5 km circuit). The first option is less exposed and easier walking; both require a good level of fitness and ability to walk in soft sand.

*Steep-sided section of track to the lake*

## Facilities

A carpark and walkers' camp (with untreated water, platform seats) are provided near the lookout. Toilets and information boards are located here and on the beach. Take your rubbish with you including cigarette butts and carry plenty of drinking water and sun protection.

## What to look for

Eleven species of native freshwater fish are believed to inhabit Lake Wabby, including catfish, the rare honey blue-eye and a species of sunfish previously not recorded south of Cairns. (Note: fishing is not permitted in any of the island's freshwater lakes). How these fish came to colonise the lake in the first place is a mystery to many, some even speculating that it may have 'rained fish' in the past!

Scribbly gums are a distinctive feature of the woodlands surrounding the lake. Some beautiful examples grow near the lookout. Lace monitors will often make

# Lake Wabby

*Lake Wabby is an impressive example of the dynamic link between sand, sea, vegetation and fresh water.*

an appearance as will torresian crows ('Wabby' is said to be the local Aboriginal word for 'crow'), while birds of prey such as whistling kites, brahminy kites and sea eagles may be spotted above, as they patrol the dunes and nearby beaches in search of their next meal.

## What we think

For many visitors, Lake Wabby is the highlight of their holiday. A few, however, return dissapointed, perhaps because of the downhill and uphill walk. But if you really want to learn about the island and get a feel for its dynamic, ever-changing nature, a visit to Lake Wabby is a must.

The view from the lookout across Hammerstone Sandblow toward the South Pacific Ocean is nothing short of spectacular. If you decide to have a closer look at the lake - especially if you're walking in from the beach - take precautions against sunburn and dehydration as the light reflecting off the sand, wind exposure and lack of shade can be severe.

Many people enjoy swimming in some of the inland lakes like Lake Wabby (although life guards are not present). If you intend to swim, avoid diving into the lake without first checking the depth. The adjacent ocean waters are unpatrolled and contain dangerous rips as well as sharks that pursue fish close to shore. This renders them unsafe for swimming.

Because it's a major drawcard, Lake Wabby can have no-one there one minute and large groups of visitors the next.

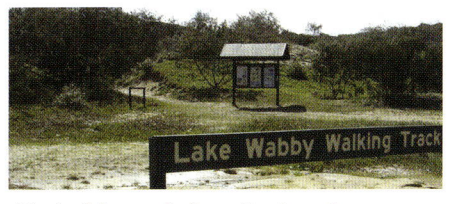

*Start of the walk from the beach*

# Southern Fraser Island

**Map reference**: S6 (Hema), L5 (Sunmap)

**Location**: 8 km from Central Station, 10 km from Dilli Village, 35 km from Hook Point

**What's there?** Seaside township and resort, information centre, beach fishing, good facilities

**Nearest to**: Central Station, Dilli Village, Lake Wabby

## Why go?

With plenty to see and do in the vicinity, the small beachfront township of Eurong is a popular stopover for many island visitors. Good facilities and a relaxed atmosphere attract people from all walks of life, from couples and families to overseas backpackers and fishing enthusiasts.

Some of the island's most popular lakes, including Wabby and McKenzie, are within an easy driving distance. So too are the stunning rainforests of Central Station and Pile Valley.

*Tourists flock to Eurong's general store*

From its comparatively humble beginnings in 1963, Eurong has steadily developed into one of the largest resorts on the island. Having recently been acquired by Kingfisher Bay Resort and Village, upgrades to the existing resort are planned over the next few years.

## Getting there

There are several options for accessing Eurong. Many drivers opt for the smoother, more direct route, approaching from the north or south along Seventy-Five Mile Beach. Plan ahead to ensure the tide is sufficiently low and conditions are favourable.

Eurong can also be reached from Central Station in the island's centre. The drive is straightforward as Eurong is clearly signposted and the tracks are one-way. Although maintained, conditions can quickly change in very dry or wet weather.

## Facilities

As well as serving as the base for the island's only taxi service, Eurong has a police station, QPWS Information Centre, resort airstrip and vehicle recovery and towing service.

Being a relatively large township, most amenities are provided including public toilets, payphones, rubbish bins, a post box and well-equipped general store with eftpos. Takeaway food, fishing gear, fuel, liquor, ice and various other provisions can be purchased.

The resort offers a range accommodation from motel units to A-frame cottages, all with kitchens or kitchenettes. It is recommended that you book ahead. Other facilities include the restaurant, bars, bistro, bakery / café, swimming pool, barbecues, tennis courts and conference facility.

Car hire, scenic bus tours and scenic flights can also be arranged. In addition, there are dozens of private holiday homes available for rent, both around the township and in adjacent Second Valley. Make sure you book ahead.

# Eurong

*After the serenity of the inland tracks and forests, Eurong emerges as a bustling, cosmopolitan centre on the island's eastern beach.*

## What to look for

The beach is a great place to take a walk or throw in a line (watch for traffic), with bream, tailor and whiting commonly caught. It is not a safe place to swim, however, due to sharks, dangerous rips and the fact that there are no lifeguards.

Between July and October, keep an eye out for humpback whales. Dolphins and sharks may be spotted, sometimes very close to shore.

Overhead, birds of prey like whistling kites and sea eagles are an impressive sight and dingoes often scavenge and hunt for food along the beach and around the township. Keep your distance and avoid feeding or encouraging them, no matter how friendly they may seem (see Dingo safety p 45).

Watch for planes as they land and take off on the beach and give them a wide berth.

## What we think

Eurong is a relaxed and cheerful settlement. Take care when crossing the road, however, as having an 'audience' can bring out the worst in a few drivers. (Unfortunately, the same goes for the other resorts and townships).

A stay here will satisfy fishing folk and visitors who like the sound of the surf - as well as their creature comforts. (However, don't expect the natural bushland setting of Kingfisher Bay Resort).

The proximity of Eurong to many of the island's most popular attractions, especially the eastern beach, makes it a sound choice for those planning a jam-packed itinerary.

# Southern Fraser Island

**Map reference**: Q5 / R5 (Hema), K4 (Sunmap)

**Location**: 9.5 km from Central Station, 14 km from Kingfisher Bay, 15 km from Eurong

**What's there?** Crystal clear water, beautiful scenery, good facilities

**Nearest to**: Basin Lake, Central Station, Pile Valley

## Why go?

Exceptional natural beauty, crystal clear waters, white sandy beaches and relatively easy access all help to make Lake McKenzie the island's most popular natural attraction. During peak holiday periods, the lake attracts up to 2000 visitors a day, many keen for a swim.

*A young dingo visits the water's edge*

Lake McKenzie is a natural wonder and one of the world's most stunning examples of a perched lake, half of which are found on Fraser Island.

Situated above the island's main water table, perched lakes are thought to consist purely of rainwater, with water lost through evaporation being replaced by rainfall. A perched lake is created when compressed sand and organic matter forms a sealed bottom layer beneath a bed of fine, pure quartz sand. The rainwater that collects is naturally acidic and remains clear with low mineral levels. Decaying organic matter releases more acids, limiting the growth of plants and algae. Bear in mind that because what goes into the lake stays in the lake, pollutants such as sunscreens, insect repellents and cigarette butts are likely to upset this natural balance.

## Getting there

Lake McKenzie is accessed from the island's eastern beach via Cornwell's Break Road or Eurong Road or from Kingfisher Bay on the western side, following signs. Because of the lake's popularity, expect to meet other vehicles en route including large tour buses which will expect you to give way. Patience and courtesy are the key and a number of one-way traffic sections make the going easier. Because of their high usage, these are some of the best-maintained tracks on the island. But they can deteriorate quickly after excessive dry or wet conditions or heavy use.

Walking to the lake is a great alternative to driving if you are relatively fit and have the time. A popular route is via Central Station and Basin Lake. Longer walks can be taken from Kingfisher Bay and Lake Wabby. These form part of the Fraser Island Great Walk. For the longer walks in particular, it's worth reading the brochure before you arrive - see p 14 -16).

## Facilities

Vehicle-based camping is no longer available (Central Station has been redeveloped as an alternative). The lake's day visitor area includes toilets, dingo-proof lockup unit, interpretive signage, picnic tables, sink and tap and extensive parking areas with separate parking and picnic facilities for bus groups. From the car park, it's a relatively short walk down to the

# Lake McKenzie

*Exceptional beauty and crystal clear waters earn Lake McKenzie 'paradise' status in the eyes of most visitors from Australia and around the world.*

main beach. The hikers' camp is fenced, with low tables.

## What to look for

Dingoes, kookaburras, lace monitors and torresian crows are commonly seen along with a small native fish known as the purple spotted gudgeon. The strange, partly submerged shapes of visiting musk ducks are observed on rare occasions.

Opportunistic dingoes and crows are attracted to rubbish and food scraps and aggressive kookaburras sometimes snatch food. Do not feed any animals and if swimming or sunbathing, keep a cautious eye out for dingoes, especially if you have young children. Resident dingoes are notorious for stealing bags, shoes, hats and other belongings. Although many people tend to be more wary of dingoes these days, incidences can still occur (see Dingo safety p 45).

## What we think

If you have limited time, this amazing lake is a must-see. No matter what the weather or how crowded the lake is, we are always awestruck by its natural beauty. However, just like Uluru, the lake has many moods and visitors should not expect a brochure replica. Early morning is great because birds sing, the water is calm and the crowds are yet to arrive. However on a calm day, the afternoon sun shining on the lake can produce some of the best colours.

Water lovers will be hard-pressed to find a more beautiful lake anywhere in the world. A beach umbrella is a good idea since sunscreen can pollute the water. Cigarette butts collect along the lake edge and do not break down so do the right thing to help keep this place special.

# Southern Fraser Island

**Map reference**: R4 (Hema), K4 (Sunmap)

**Location**: 2 km (walk) from Central Station, 4.3 km from Lake McKenzie

**What's there?** Small perched lake, freshwater turtles, scenic walks

**Nearest to**: Central Station, Lake McKenzie, Pile Valley

## Why go?

Hidden in the forest, this small and picturesque freshwater lake offers a secluded resting point for walkers between Central Station and Lake McKenzie.

Rarely described in brochures, Basin Lake often comes as a pleasant surprise to walkers not expecting to find such a pretty spot. Many choose to take a swim or quietly relax and observe the local wildlife. The lake is home to a number of Kreft's river turtles; shy creatures rarely seen at the island's busier lakes.

But what is a peaceful setting one minute, can become a busy scene the next. Daily visits by bus loads of backpackers are a sign that the popularity of this small and pretty lake is on the rise.

While the lake often appears a deep shade of blue, a closer look reveals the water is in fact crystal clear.

*Basin Lake is popular with backpackers*

Surprisingly deep, the depth of Basin Lake is said to exceed eight metres in parts. At times, the reflections cast by the surrounding forest transform the lake to a deep shade of green.

Just like its much larger neighbour, Lake McKenzie, Basin Lake is what's known as a 'perched lake'. The lake is isolated from all other water bodies, relying on rainfall to counter the effects of evaporation. Local plants and animals depend on the natural processes that keep the water clean and mildly acidic. Any upsets to this delicate chemical balance could threaten the lake's ecology and water quality.

## Getting there

Walking is the only way to get to Basin Lake. The shortest walk is from the adjacent bend in the road out of Lake McKenzie, located just west of Basin Lake. However, the start of this track is difficult to find since there are no signs or obvious parking provided.

It is advisable to take the 2 km (one way), signposted walk from Central Station. This very scenic walk starts at the Wanggoolba Creek boardwalk, passing over the creek, in the opposite direction to Pile Valley. From Basin Lake, it's another 4.3 km to Lake McKenzie. These walks form part of the Fraser Island Great Walk (see p 14 -16 for more information).

## Facilities

No facilities are provided at Basin Lake. The nearest facilities are located at Central Station.

## What to look for

Visitors often marvel at the smooth, straight white trunks of a series of planted flooded (or rose) gums located at the beginning of the walk from Central Station. Further along, it's hard to miss the soft, bright green native foxtail ferns that make

# Basin Lake

*Visitors can enjoy a beautiful walk and discover the natural beauty and ambience of this small, sheltered lake - perfect in any weather.*

up the understorey. These unusual, sand-loving plants are unique to this region.

Turtles can often be spotted in the lake, venturing very near to shore when conditions are calm and quiet. Lake Allom and Lake Bowarrady also provide opportunities for viewing these creatures but elsewhere on the island, they tend to be much more elusive. Observe quietly from the shore if you're hoping to spot some.

Kreft's river turtles are a type of freshwater, short-necked turtle, with webbed, clawed feet and the ability to fold their neck sideways if threatened by a predator. Researchers are currently trying to find out whether they are in fact a new species, different from those on the mainland.

## What we think

Proof that good things can come in small packages, Basin Lake has to be one of our favourite locations on Fraser Island. In calm weather, especially on sunny days, the water can look stunning.

However, resist any temptation to feed or encourage the turtles and take all rubbish with you, including fruit skins and cigarette butts.

If swimming, avoid wearing sunscreen or insect repellent as these will contaminate the water. Take extra care to ensure your visit doesn't spoil this very special place.

# Southern Fraser Island

**Map reference**: R5 (Hema), K4 (Sunmap)

**Location**: 9 km from Lake McKenzie, 9.5 km from Eurong, 14 km from Kingfisher Bay

**What's there?** Rainforest, clear freshwater creek, former forestry camp & plantations, QPWS office

**Nearest to**: Basin Lake, Lake Jennings, Pile Valley

## Why go?

Central Station is a magnet for visitors wanting to see the unusual phenomenon of rainforest growing in sand. Also of interest is the beautiful and much photographed Wanggoolba Creek with its crystal-clear fresh water running silently over clear, quartz sand.

Originally established as a forestry camp in 1920, Central Station is also of historical interest. Its impressive stands of timber species, which include experimental plantations, are a living reminder of the former forestry industry. Part of a thriving industry in its heyday, the logging operation went on to attract severe criticism from environmentalists until it was halted in 1991. The island received World Heritage status the following year.

*Piccabeen palms tower overhead*

## Getting there

Take the well signposted tracks from Eurong (mostly one way) or Kingfisher Bay. Although well maintained, track conditions can deteriorate quickly. An alternative route is to follow Cornwell's Break Road from the eastern beach, turning left at the sign to Central Station at the first intersection after Stonetool Sandblow. This track passes the Lake Wabby turnoff and like many other tracks, can at times be rough, narrow and overgrown in places.

## Facilities

Central Station provides excellent facilities. The day use area includes boardwalks, a QPWS Ranger Base, interpretive display and public payphone, extensive parking, picnic tables, communal gas barbecues (reserved for groups 11am-2pm), shelters, drinking water and toilets. The large fenced campground set amongst the rainforest is located just 700 metres to the east. This new $1.2 million setup can take up to 300 people. Many campsites have picnic tables, while sinks and taps are dotted throughout the campground. Toilets, coin-operated hot showers and facilities for camper trailers / vans are provided. Campsites are secured on a first come first served basis. Campfires are not permitted and there is a 9pm noise curfew.

## What to look for

Kauri pines were the island's first trees to be logged - the trees' straight trunks made excellent ship masts. Look for their heavy, cone-like fruits on the ground. Hoop pines are also easy to spot - their trunks feature hoop-like scars created when younger branches are shed.

Vertical grooves or 'striations' on the bark of the Fraser Island satinay distinguishes this species from the brush box, which has scaly bark in comparison. Look also for the planted stand of flooded (rose) gums with their beautiful straight, smooth white trunks.

# Central Station

*Below the boardwalk, crystal clear waters run silently over white quartz sand in Central Station's famous Wanggoolba Creek.*

Down along the rainforest boardwalk, one of the few remaining populations of the ancient king fern, a diminishing species Australia-wide, can be seen growing in Wangoolba Creek. Also fascinating to many visitors are the small, green patches of algae in the creek bed, their growth triggered by the filtered sunlight.

While piccabeen (bangalow) palms dominate the rainforest canopy, a variety of orchids, lianas, vines and epiphytes such as staghorns and elkhorns can also be seen. Keep an eye out for piccabeen vines (their leaves resemble animal vertebrae) as well as the giant buttressed roots of strangler figs. After beginning life in the treetops, a number of figs have grown to take on spectacular cathedral-like forms.

Rainforest wildlife is notoriously secretive, but it is not unusual to see an eel swimming up the creek or brightly coloured kingfisher perched strategically above. Dingoes, carpet pythons and lace monitors are often seen and it's hard to mistake the calls of screeching sulphur-crested cockatoos overhead or the resonating "whipcrack" of the eastern whipbird in the rainforest. Be on the lookout for kookaburras. These observant birds are famous for snatching food from the hands of unsuspecting visitors!

## What we think

Central Station's history, natural beauty and facilities make it a place well worth visiting. Be prepared, however, to share this serene location with large numbers of people when the bus groups arrive.

The best spot for taking in the sights and sounds of the rainforest and creek is along the boardwalk and walking track to Pile Valley. In nice weather, the walk to Basin Lake is well worth it for those who have the time and energy.

# Southern Fraser Island

**Map reference**: R5 (Hema), K4 (Sunmap)

**Location**: 2 km (by vehicle) from Central Station

**What's there?** Impressive satinay trees, rainforest walk to Central Station

**Nearest to**: Basin Lake, Central Station, Lake McKenzie

## Why go?

A survivor of the island's recent logging history, Pile Valley attracts visitors interested in its magnificent trees, including the famous satinay. Rainforest lovers can enjoy the very scenic walk along Wanggoolba Creek toward Central Station.

Making up less than seven percent of the island's total vegetation, the rainforest on Fraser is one of the island's claims to fame. Some of the world's tallest pockets of rainforest on sand grow here. More fascinating to scientists, however, are the insights the forests provide into long-term soil development and vegetation succession. Botanists classify Fraser Island's rainforest quite separately from other Australian rainforests. It is composed of unique combinations of species, many of which reach their most southern or northern limits.

One of Pile Valley's main attractions is the Fraser Island satinay, *Syncarpia hillii*. Although recorded for other southeast Queensland locations, this once highly sought after timber grows tallest and in greatest abundance on Fraser Island. Some of the island's best stands occur at Pile Valley; so named because the trees were used for wharf piles. Long, straight and resistant to marine borer, the timber was used to reconstruct parts of the Suez Canal and at London's Falmouth Dock. Going by the trade name 'Red Satinay', it was also considered attractive as flooring and furniture. Satinay was used in Canberra's Old Parliament House and the Prime Minister's cottage.

In 1937, the Department of Forestry declared Pile Valley a 'beauty spot'. The island was World Heritage listed in 1992, with the last of the logs harvested from the island carried off as recently as 1991. Today, nearly all of the island is protected as National Park.

## Getting there

Pile Valley is most commonly accessed from Central Station on foot (4.6 km return) or along the vehicle track between Central Station and Lake McKenzie, following signs. The first part of the drive from Central Station is one-way.

## Facilities

No facilities are provided at Pile Valley apart from an information sign. A small arrow near this sign indicates the start of the 2.4 km walk to Central Station. The

*Satinays can grow to 40 m or more*

# Pile Valley

*The walk from Pile Valley to Central Station meanders through satinay and brushbox forests down to Wanggoolba Creek and its lush surrounds.*

nearest QPWS office, public payphone, camping and other facilities are located at Central Station.

## What to look for

Satinay and brushbox dominate the rainforest, interspersed with the tall, straight trunks of hoop pines and kauri pines. Satinays are easily recognised by their deeply furrowed, fibrous bark and sometimes massive trunks. At Pile Valley, they grow up to forty metres tall. The species is a member of the Myrtaceae - the same family of plants to which eucalypts belong. Its woody, capsule-like fruits are one of the many family traits.

Carrol (*Backhausia myrtifolia*) is a common tree of the understorey and rainforest edges. When crushed, its leaves have the aroma of bubble gum. Look down as well as up because at certain times of the year, the beautiful blue fruits of blue quandongs can be seen littering the forest floor. In the creek, long-finned eels are sometimes spotted. In wet weather, this remarkable fish has been known to travel long distances over land. Although much more elusive, catfish are also worth keeping an eye out for.

## What we think

If you only do one walk while on Fraser Island, make sure it's the Pile Valley-Central Station walk. Many visitors who prefer walking downhill to uphill, start at the Pile Valley end and arrange to be picked up from Central Station.

From Pile Valley, the track leads down toward the creek bed for some distance and is then relatively flat (except for a very short incline at the end). As you approach Central Station, expect to encounter tour groups - the early and latter parts of the day are often the quietest.

# Southern Fraser Island

**Map reference**: R5 / S5 (Hema), K4 (Sunmap)

**Location**: 1.5 km from Lake Birrabeen, 6.5 km from Central Station

**What's there?** Small, secluded lake with tea-coloured water

**Nearest to**: Central Station, Lake Benaroon, Lake Birrabeen

## Why go?

A small, secluded freshwater body, Lake Jennings offers pleasant views from the road but attracts very few visitors to its interesting shores.

Sedges and rushes flourish along the lake's edge, providing habitat for insects, reptiles and the island's sensitive acid frogs. The lake's soft water, on the other hand, supports only a small amount of plant and animal life. Firetail gudgeons, a small type of freshwater fish, are able to survive in the nutrient-poor conditions.

*The curly sedge is easy to recognise*

Photographers can enjoy capturing the lake's unique beauty, with a full sun overhead making the water's tea-like colours all the more vivid. Organic acids, released by decomposing vegetation, create this special colour. The white quartz sand, which lines the bottom of the lake, prevents the acids from precipitating out.

## Getting there

When approaching from Central Station, Lake Jennings is the first lake to come into view along the Southern Lakes Tourist Drive. The track is two-way with numerous blind corners. Be prepared for the going to be slow, bumpy and potentially soft in parts.

The lake is not signposted but a little further on, the road splits with a sign indicating that the left track is for approved tour operators only. A very short walking trail to the lake's edge is situated immediately in front of this, on the left in the middle of the bollards. There is no designated parking - the only option is to pull off the road into an available space nearby. Use only existing clearings, taking care not to block tracks or damage vegetation.

## Facilities

There are no facilities. Nearby Lake Birrabeen has toilets and picnic tables.

## What to look for

A little agility is needed for the very short walk down to the water's edge as one or two fallen trees may obstruct the path. Look for the large leaning scribbly gum, banksias, orange-coloured dodder vine and pretty weeping baeckeas on either side. Take a closer look and see if you can make out the curly sedge, *Caustis recurvata* or the furry-looking foxtail, *Caustis blakei*.

Zones of vegetation become obvious as soon as you reach the shore. Sedges and rushes pioneer the water's edge, with paperbarks and sheoaks claiming the shallow soils behind. These give way to diverse eucalypt woodlands that surround much of the lake.

On the beach, tiny red sundews emerge from the sand, oozing a sticky, glistening substance specially manufactured to

# Lake Jennings

*Although often appearing a deep blue to passers by, Lake Jenning's gently lapping waters are in fact striking shades of brown and gold.*

ensnare insects.

No matter what the time of day, the surrounding forest is always alive with the sounds of birds. When the gums are in flower, the chaotic chatter of lorikeets can be heard over the melodious singing and cooing of grey shrike thrushes and bar-shouldered doves.

Smaller birds, like red backed fairy-wrens and eastern yellow robins, may be spotted flitting through the undergrowth, while musk ducks bob up and down out on the lake's surface.

Part of the Fraser Island Great Walk skirts the other side of the lake (see p 14-16).

## What we think

Lake Jennings may look enticing from the car but there are no facilities or proper carpark to accommodate visitors. Only those with a genuine fascination for ecology and willingness to 'tread lightly' ought to factor a stop into their itinerary. The small beach has just enough room to set up a couple of chairs - perfect for bird spotting, a private picnic or to catch up on your reading.

Families and groups looking for an amazing place to swim and explore, should save their stop for spectacular Lake Birrabeen, just around the corner.

*Small carnivorous sundews dot the beach*

# Southern Fraser Island

**Map reference**: S4 / S5 (Hema), L4 (Sunmap)

**Location**: 8 km from Central Station, 21.5 km from Dilli Village

**What's there?** Crystal clear waters, white sandy beach, beautiful scenery

**Nearest to**: Central Station, Lake Benaroon, Lake Jennings

## Why go?

Lake Birrabeen is one of the island's few perched lakes with crystal clear water and white sandy beaches. For many visitors, its exceptional beauty puts it in the same league as Lake McKenzie.

Like other perched lakes, Birrabeen is essentially a large pool of mildly acidic rainwater, perfectly clear and relatively free of algae and aquatic plant life. (See Lake McKenzie p 146 for more details).

*During quieter times of the year, you may have the lake to yourself*

An ideal spot to enjoy a swim or picnic in relative peace and quiet, this stunning lake attracts nature enthusiasts, water lovers, bushwalkers, backpackers, picnickers, families and a few bus groups.

## Getting there

Follow signs from Central Station or approach from Dilli Village along what is locally known as the "southern lakes" route. After leaving Central Station, Lake Jennings is the first of the main lakes, followed by Birrabeen, Benaroon and Boomanjin. Although slow-going and often bumpy, limited traffic and constantly changing scenery can make for an enjoyable journey. Conditions can deteriorate quickly, however, after prolonged dry spells or excessive rain.

The lake is also passed by bushwalkers undertaking part of the Fraser Island Great Walk (see p 14-16).

## Facilities

A small but adequate carpark is provided, leading to a nearby lake viewing platform, toilets (wheelchair access) and picnic tables. (The picnic tables are often used by bus groups). It's just a short walk down to the main beach. A second access track and picnic table are situated a short distance south. Nearest camping is available at Central Station, Lake Boomanjin or the walkers' camp at Lake Benaroon.

## What to look for

You may be surprised that the pure, clean waters of such lakes support so few life forms. Here - and in other perched lakes - a different and much simpler, but equally fascinating, foodweb has evolved.

In the lake's acidic waters, small, pretty native fish known as firetail gudgeons survive despite a lack of worms, aquatic insects, crustaceans and other food sources typical of more productive lakes. Instead, these fish feed on insects that collect on the water's surface, such as flying insects or species dependent on the lakeside habitat for parts of their life cycle.

Keep an eye out for grebes. Lobed toes make these small aquatic birds excellent swimmers and they can often be seen diving underwater for food. All three grebe species occur on Fraser Island; the hoary-headed, Australasian and great crested. Overhead, you may spot a brahminy kite or catch the piping call and yellow flash

# Lake Birrabeen

*Lake Birrabeen's crystal clear waters and white sandy beaches create an oasis on par with the beauty of Lake McKenzie.*

of an Eastern yellow robin in the forest. Binoculars and a field guide will help you to identify the many different bird species.

Dingoes are sometimes seen scouring the edges of the lake, hunting or scavenging for food. No matter how friendly or hungry they may seem, do not feed or interact with these or any other animals. Food scraps and handouts upset the area's delicate ecology and compromise the safety of visitors.

Surrounding the lake is dense vegetation, consisting of a variety of banksias, eucalypts, paperbarks and wallum heathland plants. These provide food and shelter for local animals - especially birds - as well as shade for human visitors.

## What we think

Lake Birrabeen must be one of the island's best kept 'semi' secrets. If you want the Lake McKenzie experience without the crowds, this is it. Since the two lakes are so alike, we suggest you only visit one of them if your time is very limited. Avoid the larger bus groups by visiting the lake before 11am or after 1pm.

Take care to remove all rubbish and keep sunscreens, cigarettes and other pollutants from entering the water. We all have a responsibility to look after this beautiful yet fragile place.

*Banksia in flower*

157

# Southern Fraser Island

**Map reference**: S5 (Hema), L4 (Sunmap)

**Location**: 3.5 km from Lake Birrabeen, 7.5 km from Lake Boomanjin, 11.5 km from Central Station

**What's there?** Attractive freshwater lake, turtles, hikers' camp

**Nearest to**: Lake Birrabeen, Lake Boomanjin, Lake Jennings

## Why go?

Fringed by sedges and white sandy beaches, Lake Benaroon is a medium sized perched lake situated just south of Lake Birrabeen.

Walkers enjoy the lake's wilderness values and photographic opportunities, a few opting to stay over at the hikers' camp, take a swim or simply relax on the quiet beach.

*Bright green foliage of the foxtail*

## Getting there

Drivers can view and gain access to the lake via Birrabeen Road as part of the Southern Lakes Tourist Drive - although the lake and path are not signposted. The most direct access to the water's edge is via a very short, overgrown path located behind a set of bollards. These are situated about 300 m northeast of the M3 hikers' marker and fire control line (if driving towards Lake Birrabeen), or 1.5 km southeast of the M2 hikers' marker, after Lake Barga (if driving towards Lake Boomanjin). The path brings you out onto a reedy beach on the lake's southeast shore. Tread carefully to avoid damaging the fragile plant life.

For those wishing to park close to the lake, parking space is very minimal. Use only existing clearings, taking care not to block tracks (including management and fire tracks) or damage vegetation. The nearest designated carpark is located at Lake Birrabeen, about 1.6 kilometres from the M2 marker.

The white sandy beach at the northwest end of the lake can be accessed on foot from two directions by following part of the Fraser Island Great Walk (see p 14-16). From the north, the walk passes through flowering heathlands and skirts a small, sedge-lined lagoon (Lake Barga), before reaching the shore of Lake Benaroon.

The walkers' camp is located just above the southwest corner of the beach, to the right. Alternatively, walkers can approach the beach from the southeast corner of the lake. This part of the trail hugs the edge of the lake, taking in the forest and beaches of its attractive southwest shore.

## Facilities

Lake Benaroon is not signposted for vehicles and there are no facilities. The shady walkers' camp has a dingo-proof lockup unit, toilet and platform seating.

## What to look for

Keep an eye out for firetail gudgeons, a small native fish common to many of the island's lakes. Sand yabbies - a type of freshwater crayfish - excavate burrows around the edges of the lake. Only found on the sand islands off southeast Queensland, these small, semi-aquatic creatures are specially adapted to the

158

# Lake Benaroon

*One of Fraser Island's oldest lakes, Lake Benaroon attracts just a handful of visitors who come to explore the southern lakes.*

lake's acidic water.

Kreft's river turtles also inhabit the lake but plenty of patience is needed to spot these shy, elusive creatures. They lay their eggs in burrows along the beach in an attempt to hide them from scavenging dingoes and goannas. If located and devoured, diggings and broken egg shells are often all that remains.

An impressive nighttime chorus may be heard at certain times of the year when tiny native acid frogs, like Cooloola sedgefrogs and wallum froglets, sing collectively to attract mates. Their calls, in turn, attract hungry predators like tree snakes and keelback snakes, neither of which are dangerous to humans.

Birdwatchers can have a field day spotting dozens of different species. Bar-shouldered doves feed on the ground, while robins, wrens and honeyeaters flit through the dense understorey. Paperbarks and sedges dominate the water's edge, behind which different forest types converge. Grass trees, banksias and wedding bushes spill out of the diverse heath communities, while cycads, foxtails, blackbutts and brushboxes can be spotted from the road.

## What we think

Nature enthusiasts will love exploring the interesting shoreline. However we suspect the lake's bleak, untamed beauty will be lost on some people, especially if they have just visited picture-perfect Lake Birrabeen.

If swimming or camping, take care not to upset the lake's fragile ecology by allowing sunscreens, detergents or other chemicals to pollute the water.

# Southern Fraser Island

**Map reference**: T5 (Hema), L4 (Sunmap)

**Location**: 10.5 km from Dilli Village, 11 km from Lake Birrabeen, 17 km from Central Station

**What's there?** World's largest perched lake, tea-coloured water

**Nearest to**: Lake Benaroon, Lake Birrabeen, Dilli Village

## Why go?

Impressive in terms of its size and colour, Lake Boomanjin is a popular stop along the island's southern lakes route. The lake covers an area of more than two hundred hectares (two square km) making it the world's largest perched lake.

Movie buffs will be interested to learn that this was the filming site of the Australian film 'Eliza Fraser'. Today, the windswept beauty of this lake continues to attract nature lovers, bushwalkers and families, as well as the occasional bus group.

This is also a photographer's paradise. The strongly tea-coloured water laps gently onto the white sandy beach while reeds and sedges cast intricate lines, shadows and reflections.

*In warm weather, visitors come for a dip*

## Getting there

Follow signs from Central Station or approach from Dilli Village along what is locally known as the 'southern lakes' route. After leaving Dilli Village, Lake Boomanjin is the first major lake, followed by Lake Benaroon, Lake Birrabeen and Lake Jennings. Although slow-going and often bumpy, limited traffic and constantly changing scenery make for an enjoyable journey. Track conditions, however, can deteriorate quickly after prolonged dry spells, excessive rain or heavy use.

The lake is one of the highlights of the Fraser Island Great Walk (see p 14-16). Bushwalkers can approach from Dilli Village or Lake Benaroon.

## Facilities

Parking, toilet and picnic facilities are provided. The camping area overlooks the lake and is enclosed by a dingo-proof fence. Campsites cannot be booked in advance. Showers and drinking water are provided but campfires are not permitted.

## What to look for

A common plant is the paperbark teatree with its gnarled, twisted trunks and strong-smelling foliage. Aborigines made good use of its light, fibrous bark for constructing shelters, lining their canoes and cooking fish. The leaves were valued for their antiseptic qualities.

Numerous bottlebrush-like flowers produce large amounts of nectar, attracting birds such as honeyeaters and lorikeets and a range of other wildlife including native bees, bats and sugar gliders.

Dingoes are sometimes seen as are the few native frog and fish species able to tolerate the lake's acidic environment. Whereas most of the island's other lakes support only one species of fish, Lake Boomanjin is home to native rainbow fish, firetail gudgeons and purple spotted gudgeons.

The lake is often at its most photogenic

# Lake Boomanjin

*Lake Boomanjin's tea-coloured waters lap gently onto its white, sandy shoreline making the world's largest perched lake a striking and beautiful location.*

early morning or late afternoon when the air and water are still and the angle of light is low.

Tannin - the same chemical present in tea - gives the lake's water its distinctive colour. Released by decaying vegetation, the tannins remain in the water because they are unable to react with the pure quartz sand lining the bottom of the lake. (Coloured sand reacts with the tannins and produces clearer water).

## What we think

Once you've seen one perched lake you've seen them all? Not at all. Across the island, differences in size, appearance and even water colour give each lake its own unique character so if you have the time, Lake Boomanjin is definitely worth a look. It is an essential stop along the 'southern lakes' route and is one of our favourite picnic spots.

Although the campground is fenced and offers protection from dingoes, it looks a little sparse and cage-like. We prefer a less enclosed outlook.

Out of the peak tourist season, this part of the island is generally quiet, attracting a few like-minded visitors looking to get off the main drag and escape the crowds. If this sounds like you, we recommend setting aside at least a day to fully explore the southern lakes.

*Spectacular tea-coloured water*

# Southern Fraser Island

**Map reference**: U5 (Hema), M5 (Sunmap)

**Location**: 10 km from Eurong, 10.5 km from Lake Boomanjin, 24 km from Hook Point

**What's there?** Grassy camping, cabins, eco-education site, water hole

**Nearest to**: Eurong, Lake Benaroon, Lake Boomanjin

## Why go?

Dilli Village provides a convenient base for those wanting to explore the island's southern lakes region on either foot or four wheels. In addition to the newly renovated budget accommodation, visitors have the option of well-grassed campsites which offer plenty of room for family groups and camper trailers.

Now managed by the University of the Sunshine Coast, Dilli Village doubles as a unique eco-educational site for students and special interest groups.

Constructed as a sandmining base in 1974, in conjunction with a nearby air strip and Dillinghams Road, Dilli Village derives its name from the large American industrial company, Dillingham. Together with Murphyores, the mining consortium's aspirations sparked many years of public outcry against plans to extract mineral sand from thousands of hectares of pristine dunes. Led by conservationist John Sinclair and the Fraser Island Defenders Organisation, the fight was eventually taken to the High Court of Australia and in 1976, the Australian Government banned the export of minerals from the island.

Today, the site is managed as both a public campsite and environmental education camp for secondary and tertiary students. The university encourages families, day visitors, small tour groups and environmental study groups to use the site.

## Getting there

Dilli Village is located just off Seventy-Five Mile Beach, south of Eurong. The beach turnoff is signposted and can at times be very soft, requiring high revs and a little momentum. Aim to drive on the beach as near as possible to low tide, particularly when approaching from the south (see Hook Point p 172-173). Dilli Village can also be approached from Lake Boomanjin or from the west coast along Dillinghams Road.

## Facilities

The budget accommodation options include five self-contained cabins (sleeping four-five) and 20 bunk-style dormitory rooms (sleeping two each). Extensive makeovers (including new beds and mattresses) have brought the original cabins and bunk-rooms up to date.

Campers can utilise the spacious grassy campsites, washing machine, toilet and shower amenities. Other facilities include picnic and barbecue areas. Note: since the resort is now privately run, camping is not covered by the QPWS camping fee. There is also a vehicle day use charge of $10. Be sure to take all rubbish away with you.

## What to look for

Swimmers are often drawn to the small freshwater lake located nearby. Known as the Dilli Village swimming hole, this naturally tea-coloured water body is encircled by pandanus palms and sedges. Another pretty spot is Gerowweea Creek, which flows parallel to the beach and can be viewed just 1 kilometre north of Dilli Village. Aquatic plants flourish in the creek, including some magnificent water lilies, so keep your camera handy. While it is possible to drive up alongside the creek, camping is not permitted.

Dilli Village often serves as a starting (or

# Dilli Village

*Dilli Village offers pleasant, 'no frills' camping for those who want to avoid the comparative hustle and bustle of the other managed campgrounds.*

finishing) point for bushwalkers. The 6.3 km walk to Lake Boomanjin forms part of the Fraser Island Great Walk (see p 14-16). This walk crosses a boardwalk, then takes you through woodlands and some great scenery along the way, with moderate fitness required for the uphill sections. Wongi Sandblow affords some spectacular views. From Lake Boomanjin, the trail continues northwards, towards the Lake Benaroon Walkers' Camp.

The inland drive leading northwards through the 'southern lakes' region is very scenic. On the track to Lake Boomanjin, you can enjoy views of Jabiru Swamp - the largest swamplands on the eastern side of the island - as well as some beautiful eucalypt forests.

## What we think

Dilli Village is well situated for exploring the southern half of Fraser Island as well as driving to or from Hook Point. We would choose it for its convenience, however, over its overall beauty.

Given the improvements that the university is making to the site and who they are catering to, we imagine Dilli will suit the budget and needs of some visitors perfectly.

*Swimmers love the little water hole*

163

# Southern Fraser Island

**Map reference**: S3 (Hema), L3 (Sunmap)

**Location**: 9.5 km from Central Station, 19.5 km Dilli Village

**What's there?** Quiet, calm beach, fishing, camping with views to islands

**Nearest to**: Basin Lake, Central Station, Lake Jennings

## Why go?

Tucked away on the sheltered south-west shores of the island, Ungowa offers a quiet, waterfront getaway for anchoring a boat, fishing, camping or simply enjoying a leisurely picnic in pretty surroundings.

Here, you can watch the sun set over the calm waters of Great Sandy Straits and escape the sound of revving vehicles; qualities which make this ex-forestry camp a pleasant day-trip or camping alternative to the eastern surf beach.

*Bungwall fern was an important food source for the Butchulla people*

## Getting there

From the Wanggoolba Creek barge landing, head south towards Garry's Anchorage, first heading inland along Wanggoolba Road for 2.5 km then turning right onto Ungowa Road for a further 6.5 km, veering to the right near the end.

Ungowa can also be accessed by driving across the island from Eurong or Dilli Village or by skirting the south-west coast along the slow-going Southern Road (see Hook Point). From Dilli Village, the first 7 kms inland is generally soft and bumpy. Continue heading westwards along Dillinghams Road (do not take the Lake Boomanjin turnoff on the right). The next 10 km is relatively straight and well maintained. But for these reasons, be wary of oncoming traffic travelling at high speed, particularly around corners. Make sure you keep under the 35 kph speed limit. Take the Ungowa turnoff on the right for the last two kilometres.

## Facilities

A small grassed camping and picnic area overlook the water, surrounded by pine trees and a large fig tree. (Before pitching your tent you will probably want to move the pine cones out of the way!) Two picnic tables are provided.

A dingo-proof lockup unit is provided to protect food and rubbish and toilets are located near to the ramp. Drinking water is available but rubbish bins are not, so you will need to take all rubbish with you. Boat anchorage is located a short distance to the south.

## What to look for

Several small islands are visible from Ungowa, the largest being Turkey Island. See if you can spot any isolated trees growing on the sandy spits - these 'grey mangroves' have the ability to survive in salt water. The calm waters of the Great

# Ungowa

*Once a forestry camp, Ungowa offers campers and day visitors stunning, uninterrupted views of the Great Sandy Straits and nearby islands.*

Sandy Straits are alive with animal life so keep an eye out for dolphins, dugong, sea birds and even schools of small fish leaping out of the water.

Biting insects such as mosquitoes and sand flies may also pay you a visit. The nearby mangroves and sandflats are essential to many of Fraser Island's plants and animals. A diversity of native plants can be observed around the campsite, including swamp banksias, blueberry ash and bungwall ferns. When prepared in the traditional way, the roots of bungwall ferns were an important source of carbohydrate for local Aborigines.

In the vicinity are two wrecks - the Ceratodus and Palmer. At high tide, there is virtually no beach to speak of and although a feature of interest, the condemned jetty is potentially dangerous and strictly off limits.

At night, dingoes routinely check the area for food, rubbish and loose belongings so be sure to use the lockup facility provided or somewhere secure like the back of your vehicle. If the gums are in flower, the night might be filled with the sounds of visiting fruit bats, squabbling and calling noisily from the trees above.

## What we think

Ungowa feels like the place that time forgot, made redundant when its activities as forestry headquarters (beginning in 1959) came to a grinding halt in 1992. However, it is enjoying a happy retirement as a very pleasant place for unhurried visitors to camp, fish or picnic and has real potential to become a historical point of interest. Its stunning views across the straits make it the perfect place to relax and quietly watch the sun go down.

# Southern Fraser Island

**Map reference**: V3 (Hema), M3 (Sunmap)

**Location**: 17 km from Ungowa, 31.5 km from Dilli Village, 40 km from Hook Point

**What's there?** Moored boats, fishing, sheltered clearing on the waterfront

**Nearest to**: Snout Point, Southern Track, Ungowa

## Why go?

With its safe, sheltered moorings, pleasant outlook and good fishing, Garry's Anchorage or 'Garry's Camp' is a popular island stopover for boaties and fishing enthusiasts.

The camp was named after Garry Owens; a local Aborigine who was well-known for his abilities as a tracker (although Garry is thought to have been an Anglicanised version of the Aboriginal word K'gari or Gurree).

*Information board at Garry's Anchorage*

## Getting there

While most visitors access this spot by boat, a few come by vehicle or on foot. For island travellers, there is the option of two tracks but only Southern Road is maintained. The other is a minor track and has the potential to be overgrown in parts - even obstructed - and very slow-going. Southern Road begins to the north near South White Cliffs. The minor track begins at Hook Point to the south. The journey from Hook Point can be especially slow-going but is very scenic (see Snout Point p168-169 and Southern Track p170-171). Small branches occasionally obstruct the track. If you decide to drive over them, take care if your vehicle has lowered tyre pressures.

As you approach Garry's Anchorage from the north, the last part of the track is elevated, affording some great glimpses of the water through the trees.

Note that beach driving is not possible anywhere between Coolooloi Creek and Moon Point. The island's western beaches are much narrower than the ocean beaches and often contain mangrove mud.

## Facilities

Boat anchorage, campsites, toilets, an undercover picnic table and rainwater (not suitable for drinking) are the only facilities provided. The camping area is fairly large, potentially accommodating around 8-10 camping units, with flat sites on compacted sand and leaf litter. A very interesting information board is provided that is geared towards the boaties. It includes a map highlighting areas of significance such as protected dugong habitat and a photo depicting the damage that irresponsible boat users can cause to seagrass beds.

## What to look for

Dugongs can occasionally be spotted along the island's south-west coast, appearing as large, slow-moving shapes that only surface once every five to ten minutes. Between here and Cooktown, the total number of dugongs is thought to have halved since 1980. Sadly, only a few thousand may be left on the planet.

# Garry's Anchorage

*In calm weather, Garry's Anchorage treats its visitors to all the beauty that Fraser's west coast has to offer.*

These fascinating mammals can live up to 70 years, but their reproductive rates are low.

Dolphins can often be spotted and the mangroves support a variety of native animals, including plenty of fish and birds as well as Queensland's famous mud crab. Be prepared for lots of mosquitoes and sandflies though.

The small wetland at nearby Lake Garry is not easily accessible because of the dense vegetation.

## What we think

Garry's Anchorage generally doesn't find its way into the itineraries of most land travellers but is worth a look if you are in this neck of the woods. It's a pretty little spot and ideal if you're on a boat. Campers may prefer the privacy of Ungowa, however, which has the same west coast scenery but is more centrally located.

*Mangrove seedling*

# Southern Fraser Island

**Map reference**: W3 (Hema), N3 (Sunmap)

**Location**: 3 km from Southern Track, 16 km from Garry's Anchorage, 30 km from Hook Point

**What's there?** Waterfront campsite, small beach, cabbage palms, fishing

**Nearest to**: Garry's Anchorage, Hook Point, Southern Track

## Why go?

Snout Point is the pocket-size embodiment of all that is beautiful and unspoiled about the island's west coast. Campers, anglers, nature lovers and photographers are most likely to be captivated by its charms and it's a great spot for a quick break on your way to or from Hook Point.

The small campsite is tucked in behind a little white sandy beach, affording visitors uninterrupted views across the calm waters of the Great Sandy Straits.

Fallen trees protrude from the water at high tide, one very large one providing interesting photographic opportunities. At low tide, the beach connects with the emerging sand flats, making it possible to take long leisurely walks.

## Getting there

Snout Point can be reached by vehicle by taking the minor inland track along the island's southwest coast. (The beach is impassable between Coolooloi Creek and Moon Point). This track can be overgrown in parts and is generally not maintained. In unfavourable weather, track conditions can deteriorate quickly, creating difficult or impassable sections (eg fallen trees). For this reason, this track is not for anyone in a hurry. Be prepared for the unexpected and for the going to be potentially slow.

Approaching from the south, the drive is very scenic and elevated in parts, with spectacular glimpses across the Great Sandy Straits (see Southern Track). Drivers heading south from Garry's Anchorage need to apply extra caution - you may encounter small fallen branches that have the potential to stake and consequently damage your tyres if they are highly deflated. Keep an eye out for the beautiful low swampland on the right, and shortly after the small timber creek crossing, you will begin to see your first cabbage palms.

The turnoff to Snout Point is signposted. While this begins as an open white sandy track, it soon becomes overgrown with dense, low overhanging vegetation. Dead twigs and branches may brush up against the sides and top of your vehicle.

## Facilities

There are no facilities at Snout Point apart from a clearing big enough to accommo-

*Flowering shrubs spill onto the beach*

# Snout Point

*Framed by a lush forest of cabbage palms, Snout Point has to be one of the most stunning little beaches on the island's southwest coast.*

date one camping party only. Campfires are not permitted.

## What to look for

The most striking thing about Snout Point is the cabbage palms, *Livistona australis*. While they only occur in a few isolated patches on Fraser Island, the same species grows on the coast of the adjacent mainland and places like South Stradbroke Island. In some unusual locations, like Palm Valley in central Australia, the changing climate caused the rainforests to shrink into a few remaining pockets of 'refugia'. In south Queensland, however, landclearing has reduced much of the palm forests' original range.

At low or high tide, there are plenty of mangroves to be discovered a little way along the beach. The aerial roots of the grey mangrove are easy to spot, appearing as multitudes of tiny snorkels protruding from the ground. Bird lovers can enjoy the diverse birdlife supported by the different types of habitat, including the large numbers of migratory birds that make the long journey to these sand flats each year. All up, more than 300 species have been recorded for the whole of the island so take your binoculars and field guide.

Note that the mangroves and palm forest support a formidable mosquito population. If you're prepared to share their environment with them, why not try the cooler months or bring a mozzie-proof tent?

## What we think

Boaties and island travelers - especially antisocial campers - can find a little touch of paradise at Snout Point. With its natural tranquillity, stunning scenery and sense of isolation, this unique spot offers the perfect island escape - or at least it would do if it weren't for the ferocious mozzies!

# Southern Fraser Island

**Map reference**: W4 / X4 / Y4 / Y5 (Hema), N4,O4 (Sunmap)

**Location**: 1 km from Hook Point, 13 km from Garry's Anchorage

**What's there?** Stunning west coast scenery, interesting & varied drive

**Nearest to**: Garry's Anchorage, Hook Point, Snout Point

## Why go?

The section of track roughly between Hook Point and Tootawwah Creek is unlike any other on the island, attracting the more 'seasoned' Fraser Island visitor who is looking for something different.

Since it's not possible to drive on the beach between Coolooloi Creek and Moon Point, this track provides visitors with one of the few means of taking in the island's stunning west coast scenery. Much of the drive is elevated and varied as you travel up and down dunes and meander along ridgetops, the birds-eye view making for a very scenic drive. The going can be rough though, with sections of the track quite bumpy and overgrown.

Traffic is usually minimal, plant and animal life is plentiful and the views across the calm waters of the Great Sandy Straits vary from pleasant glimpses to superb panoramas.

## Getting there

This drive involves following the continuation of Southern Road - an inland track along the island's southwest coast. It can be undertaken from either the direction of Garry's Anchorage to the north or Hook Point to the south. The most scenic section (described here) is the 26km between Hook Point and Tootawwah Creek.

This track is generally not maintained and may therefore be overgrown in parts. Unfavourable weather such as strong winds or very dry or wet conditions can cause track conditions to deteriorate quickly, creating potentially difficult or impassable sections (eg fallen trees). It is not uncommon for small fallen branches to lie across the track. Most drivers choose to drive over those that cannot be moved to the side. This may, however, present problems for vehicles with highly-deflated tyres or insufficient clearance. Like other minor tracks on the island, drivers are advised to take care and be prepared for the unexpected and for the going to be potentially slow. When passable, it may take a few hours to reach Garry's Anchorage from Hook Point, for instance.

Please note that this track is not an official scenic drive and is therefore not depicted as such on maps or in other guidebooks. At times, the track may be closed so

*Woombye in flower*

# Southern Track

*Wildlife is abundant and with a different scene around every bend, this drive is a slow but great way to enjoy the stunning scenery of the south-west coast.*

check the latest beach and track report issued by QPWS before departing.

## Facilities
There are no facilities located anywhere along this section of track. Garry's Anchorage has toilets and picnic tables but the nearest settlement is Dilli Village on the other side of the island.

## What to look for
The drive becomes very elevated in parts, providing spectacular views across the Great Sandy Straits. As you get nearer to Hook Point, it is possible to see the imposing coloured dunes of Rainbow Beach on the mainland.

The vegetation constantly changes. On top of the dunes, bracken ferns and cycads dominate the understorey, in contrast to the beautiful swamplands that flourish below. Cabbage palms begin to appear as you approach Tootawwah

Creek and the turnoff to Snout Point.

There is plenty of wildlife worth keeping an eye out for thanks to the variety of habitat types. Dingoes, for instance, may be spotted scent marking, socialising or foraging for food. If you're lucky, you could even catch a glimpse of a mother with her pups. In all instances, it is best to quietly observe from the car.

## What we think
The first time we travelled this route, we knew we'd stumbled on something special - that's why we decided to assign it unofficial 'scenic drive' status (and have arbitrarily called it 'Southern Track'). While it certainly isn't for most people, we like to think of this little-known section of track as the island's own 'Great Ocean Road'. If you're departing via Hook Point - and have plenty of time - the magnificent views can be a great way to soften the blow of your reluctant departure.

# Southern Fraser Island

**Map reference**: Y4 / Y5 (Hema), O4 (Sunmap)

**Location**: 25.5 km from Dilli Village, 35.5 km from Eurong

**What's there?** Wild beaches, fishing

**Nearest to**: Garry's Anchorage, Snout Point, Southern Track

## Why go?

From the moment you drive off the barge and onto the beach at Hook Point the adventure begins. Around to the left are the relatively calm waters of Tin Can Bay, overlooked by lush cabbage palms. To the right, the roar of the ocean and one of the roughest roads on the island.

Hook Point is the island's closest point to the mainland and a popular arrival and departure point. It's an area where the forces of nature - wind, rain, currents and waves - continually rework the sand.

*The old mining road delivers a rough ride*

## Getting there

On the mainland, the Inskip Point ferry landing can be reached by driving up the beach from Rainbow Beach, providing the tide is sufficiently low and beach conditions permit. Alternatively, you can take the bitumen road that begins near the Rainbow Beach township. Vehicular ferries depart relatively often, depending on demand and take approximately 15 minutes to land just west of Hook Point on Fraser Island (see p 17-18 for operating times and booking details).

If you hope to head north up the island's eastern beach, plan to set off as near to low tide as possible. In poor weather conditions and nearer high tide, the beach usually becomes impassable. Seawater can collect in lagoons and temporary pools, large branches can wash ashore and waves may reach high onto the beach, even at low tide. Logs, branches and other debris are a common sight, even in normal conditions. On a falling tide, a few drivers choose to proceed cautiously, regularly stopping to study the waves. Most, however, choose to boil the billy and wait it out. Further along, take care when crossing creeks (see Creek crossings p 26-27). The size, banks and course of the creeks change regularly.

If the beach appears untrafficable, either wait for suitable conditions or take the alternative inland route towards Dilli Village (signposted to Eurong). Although composed of gravel and bitumen to Tooloora Creek, a large section of this old mining road is in an extreme state of disrepair. Most visitors find the going very slow and uncomfortable because of the bumps and potholes. Tide permitting, most drivers heading north return to the beach just after Tooloora Creek.

From Hook Point, a bumpy, overgrown track meanders up the western side of the island toward Snout Point and Garry's Anchorage (see p 170-171).

## Facilities

There are no facilities at Hook Point, apart from signs located on the beach for the barge companies. Visitors wanting to return to the mainland simply wait on

# Hook Point

*Hook Point affords great views and a short barge journey to and from the mainland.*

the beach during barge operating hours. Barges come across when ready. Coolooloi Creek has no facilities but there are a few clearings for camping.

## What to look for

At Coolooloi Creek, the beach is sheltered and suitable for fishing plus there are quite a few bush campsites. Biting insects, however, can be a problem, especially in the absence of the usual southeasterly breeze. Tall cabbage palms grow in the vicinity and there are pleasant views across Tin Can Bay towards the mainland. (To drive to Coolooloi Creek from the beach, turn left at the T-junction and follow the sealed road for 3 km. Note the beach at Coolooloi Creek can be soft and potentially hazardous for vehicles).

On the opposite side of the island, Jabiru Swamp runs parallel to the old mining road. This extensive wetland is difficult to appreciate from the car. Bird enthusiasts, in particular, might enjoy a closer look. Large wetland birds such as brolgas are occasionally seen.

## What we think

The Inskip - Hook Point crossing is an enjoyable and interesting option when traveling north to Fraser Island via Noosa The drive up Rainbow Beach is spectacular. Those planning to depart the island from here should give themselves plenty of time. When forced to use the old mining road, visitors are often shocked by its poor condition and arrive at their destination much later than expected.

When it comes time to make your reluctant passage back to the mainland, the magnificent views towards Rainbow Beach may offer some consolation.

*Tooloora Creek, Seventy-Five Mile Beach*

# Further reading

For those interested in finding out more about Fraser Island a number of informative books have been written. Covering different aspects of the island's history, environment, fishing, cultural significance and other topics, these include:

Barker, J., Grigg, G.C. and Tyler, M.J. 1995. A Field Guide to Australian Frogs. Surrey Beatty and Sons, Chipping Norton.

Buchanan, R. & H. 1996. Fraser Island and Cooloola Visitors Guide. Hema Maps, Logan City QLD. (out of print)

Eckert, D. 2003. A History of Forestry & Timber: Maryborough / Fraser Island. Maryborough QLD.

Harmon-Price, P. 1995. Fraser Island, World Heritage Area: Treasures in the Sand. Dept of Environment and Heritage, Brisbane QLD.

McCarthy, B. 2003. Fraser Island: The Essential Visitors Guide. Dirty Weekends Australia, Northgate QLD.

Simpson, K. and Day, N. 1999. Field guide to the birds of Australia 6th edition. Viking Penguin, Chatswood NSW.

Sinclair, J. 1990. Fraser Island and Cooloola. Weldon Publishing, Sydney NSW.

Wildlife of Greater Brisbane. 1995. Queensland Museum, Brisbane QLD.

Williams, F. 2002. Princess K'Gari's Fraser Island. Fred Williams Enterprises.

Williams, F. 1982. Written in Sand: A History of Fraser Island. Jacaranda Press, Milton QLD.

Van Driesum, R. 2004. Discover Fraser Island. Hema Maps, Eight Mile Plains, QLD.

Useful websites include:

Australian Government Department of the Environment and Heritage- www.deh.gov.au/heritage/worldheritage/sites/fraser/

Fraser Coast South Burnett Regional Tourism Board - www.frasercoastholidays.com.au

Fraser Island Defenders Organization - www.fido.org.au

Friends of Fraser Island Inc. - www.friendsoffraserisland.org

Great Sandy Publications - www.greatsandy.com.au

Queensland Government Environmental Protection Agency / Queensland Parks and Wildlife Service - www.epa.qld.gov.au

www.greatsandy.com.au